C000139646

# 1,000,000 Books

are available to read at

---◆---

# www.ForgottenBooks.com

---◆---

### Read online
### Download PDF
### Purchase in print

ISBN 978-1-5279-0762-1
PIBN 10913529

This book is a reproduction of an important historical work. Forgotten Books uses
state-of-the-art technology to digitally reconstruct the work, preserving the original format
whilst repairing imperfections present in the aged copy. In rare cases, an imperfection in
the original, such as a blemish or missing page, may be replicated in our edition. We do,
however, repair the vast majority of imperfections successfully; any imperfections that
remain are intentionally left to preserve the state of such historical works.

Forgotten Books is a registered trademark of FB &c Ltd.
Copyright © 2018 FB &c Ltd.
FB &c Ltd, Dalton House, 60 Windsor Avenue, London, SW19 2RR.
Company number 08720141. Registered in England and Wales.

For support please visit www.forgottenbooks.com

# 1 MONTH OF
# FREE
# READING

## at

## www.ForgottenBooks.com

By purchasing this book you are eligible for one month membership to ForgottenBooks.com, giving you unlimited access to our entire collection of over 1,000,000 titles via our web site and mobile apps.

To claim your free month visit:
www.forgottenbooks.com/free913529

* Offer is valid for 45 days from date of purchase. Terms and conditions apply.

English
Français
Deutsche
Italiano
Español
Português

# www.forgottenbooks.com

**Mythology** Photography **Fiction**
Fishing Christianity **Art** Cooking
Essays Buddhism Freemasonry
Medicine **Biology** Music **Ancient
Egypt** Evolution Carpentry Physics
Dance Geology **Mathematics** Fitness
Shakespeare **Folklore** Yoga Marketing
**Confidence** Immortality Biographies
Poetry **Psychology** Witchcraft
Electronics Chemistry History **Law**
Accounting **Philosophy** Anthropology
Alchemy Drama Quantum Mechanics
Atheism Sexual Health **Ancient History**
**Entrepreneurship** Languages Sport
Paleontology Needlework Islam
**Metaphysics** Investment Archaeology
Parenting Statistics Criminology
**Motivational**

HARVARD UNIVERSITY.

LIBRARY

OF THE

MUSEUM OF COMPARATIVE ZOÖLOGY.

26285.

GIFT OF

Samuel Henshaw.

October 27, 1905.

**Proceedings of the Boston Society of Natural History.**

Vol. 32, No. 6,

p. 145–202, pl. 12–17.

---

# ALIMENTARY CANAL OF THE MOSQUITO.

By Millett T. Thompson.

Compliments of

MILLETT T. THOMPSON,

Instructor in Biology,

Collegiate Department, Clark University,

Worcester Mass.

# No. 6.— ALIMENTARY CANAL OF THE MOSQUITO.[1]

## BY MILLETT T. THOMPSON.

### INTRODUCTION.

LITERATURE on the mosquitoes is rather extensive, especially that which deals with the genus Anopheles. Comparatively few articles, however, discuss the internal anatomy of these flies and in many of those papers where details of internal structure are given, these are not cited from morphological motive. The structure of the salivary glands or the alimentary canal may be described, but it is to make clear the life history of the malarial Plasmodium or as a guide in dissecting when studying these parasites, and not from interest in the anatomy itself. In view of the secondary purpose of such descriptions it has not seemed best in this paper to make a complete survey of articles of this class or to comment at length on the interpretations of structure that are presented. Criticism and discussion will be limited to the few researches which deal extensively with the internal anatomy or which discuss this from a morphological view-point.

Of the studies that fall into this last group, several are found which give accounts of Anopheles. Nuttall and Shipley's, "The structure and biology of Anopheles" and Christophers' "The anatomy and histology of the adult female mosquito" are comprehensive and very valuable. Annett and Dutton's "Report" is excellent but more limited in scope. As far as anatomical details are concerned at least, Giles' "Handbook of the gnats or mosquitoes" is unreliable. I have been unable to obtain Grandpré and Charmoy's "Les moustiques." With respect to Culex on the other hand, I know of no single comprehensive work, but scattered descriptions of the various internal organs exist. The external form of the larvae and pupae of mosquitoes has been studied considerably, but the internal anatomy of these stages has been neglected. Raschke gives the best account for the larva in "Die larve von *Culex nemorosus*." Our knowledge of the metamorphosis of the internal organs

[1] From the Biological laboratory, Clark university, and the Laboratory of the U. S. bureau of fisheries, Woods Hole, Mass.

is summed up in two excellent papers by Hurst, "The pupal stage of Culex" and "The post-embryonic development of a gnat," and in details given by Miall and Hammond ('92) from Hurst's unpublished notes.

Three species of Culex as they occurred at different seasons furnished the material on which this paper is based. *C. stimulans* and *C. pipiens* were used for the study of the perfect insect, and a small midsummer form that was not identified gave me larvae and pupae. Comparative notes were made at all stages, but failed to bring out appreciable differences between the three forms at any time. The genus Culex was used instead of the more important Anopheles, because early in the study it became impossible to obtain a sufficient number of specimens of the latter, either larvae or adults. However, enough material was available to enable me to compare the two genera with respect to many structures and to check the descriptions given in the literature dealing exclusively with Anopheles. The specimens of Anopheles at my disposal were representatives of the commonest New England species, *A. punctipennis*.

### TECHNIQUE.

The chitin of the imago of the mosquito is not heavy enough to offer especial obstacles to research, except as it interferes with the penetration of the killing fluids. This was overcome without injury to important structures by cutting off the dorsum of the thorax while the insect was immersed in the reagent. Of all killing fluids that were tried, Gilson's mercuric nitrate proved most valuable, and in the end was exclusively used. In order to remove air from the scales, the mosquito was dipped for a moment in alcohol before it was plunged into the warm fluid. I made Gilson's fluid according to the following formula: —

| | | |
|---|---|---|
| 70 % alcohol | 10 | parts. |
| Distilled water | 86 | " |
| Corrosive sublimate (crystals) | 2 | " |
| Glacial acetic acid | ½ | |
| Nitric acid (80 %) | 1½ | |

Dissection is of primary importance in a study of the thoracic and abdominal portions of the alimentary canal of the mosquito. But it must be constantly controlled and supplemented by sections. Knowledge of the structures within the head of the insect on the other hand, has to be derived almost wholly from sections. Reconstruction proved of slight value. I used serial sections cut in the three planes usually employed and depended on careful study of these, adding to and controlling the interpretations by study of dissections and of material cleared with caustic. Excellent thick sections — 30 $\mu$ or over — for control and demonstration were prepared by killing in Flemming's fluid and allowing the specimens to blacken somewhat. Material treated thus may be imbedded in paraffin, sectioned, and the sections mounted without staining. The discoloration produced by the killing fluid gives sufficient character to the structures.

With the larva of the mosquito, study of the living animal or of fresh dissections is of paramount importance. Sections of the head especially are not readily interpretable taken alone. Fortunately for research, the wriggler of Culex has a large, transparent head, so that the contained structures can easily be made out in the living animal or in whole mounts of the head. In the light of knowledge obtained in this way, sections become interpretable and through these in turn we are enabled to understand sections of the Anopheles wriggler, where the head is small and opaque. An excellent method for mounting the whole head is to stain with picro-carmine and then clear through Weigert's fluid. Fine preparations may also be obtained by staining with haematoxylin, but this method is slower and less uniformly successful than the picro-carmine stain.

The pupa stage can only be studied from serial sections and to work to best advantage a series of specimens the ages of which are approximately known, is needed. Such a series was obtained for Culex by segregating mature larvae in a dish and each hour removing all pupae to separate containers, in which they could be reared for any desired number of hours. A very complete series ought to be obtained. In the summer our species of Culex spend from 48 to 70 hours as pupae, and I did not find that a set of specimens representing in ages nearly every hour up to the thirtieth of pupal life and then more distant intervals, was too extensive.

## Imago.

*Head and mouthparts.* — An extended description of the external anatomy of the mosquito does not come within the scope of this paper. A few points with respect to the head, however, must be noted. The part of the head that lies in front of the large compound eyes is inflated above to form a rectangular box, which is called the clypeus. This seems to correspond to the "face" of other flies. A vertical furrow is impressed into the anterior face of this box. At the foot of the furrow a boss of chitin for muscle insertion projects into the cavity of the clypeus, while near the roof of the box on either side a short chitinous ala also enters the cavity. Ventrally and at the sides, the pre-ocular region of the head is rounded and forms as it were an imperfect cylinder, at the summit of which the mouthparts arise, nearly at one level. The anterior wall of the clypeus lies forward of this point by one half of the length of the box in the female insect and nearly two thirds of the length in the male. The postocular or epicranial region of the head is extensive in the female but truncate in the male mosquito.

Within, the head is strengthened by a mesial crest which continues from the level of the antennae along the roof of the head to the border of the occipital foramen. It is also braced by two hollow cylindrical struts which pass from the ventral border of the occipital foramen to the cheeks. I shall refer to these struts as tentoria, without intending to imply any necessary morphological connection between them and the analogous internal head braces of other insects. Among the flies, hollow tentoria similar to those of the mosquito are found in Chironomus, Anopheles, and Simulium. They appear to be wanting in many families, as, for example, the Tipulidae, Asilidae, Dolichopodidae, and the Muscidea. The Tabanid flies have solid tentoria with somewhat the same relations as the hollow struts of the mosquito. In the female of Culex the tentoria arise in front of the border of the occipital foramen and ascend at an angle of twenty-five degrees with the floor of the head. Each strut has a short spur near the lower end, to which no obvious function can be assigned, and above as the cheeks are approached, flares out into an irregular funnel-shaped "head." In the male mosquito the struts arise from the border of the occipital foramen, and the

differently shaped head makes the angle with the floor measure
nearly thirty-five degrees.  The struts are provided with the seemingly useless spur near the foot, but above expand suddenly to form
the "head" and then the tube narrows again as it meets the cheeks.

The mouthparts (pl. 16, fig. 46–47) of the mosquito need no
elaborate description here.  The usually accepted nomenclature will
be employed for the various stylets.  The labrum is horseshoe-
shaped in section and forms the whole or the major part of the tube
(pc) through which the insect sucks blood or other liquid.  There is
no good reason to retain the name labrum-epipharynx for this dorsal stylet, with the consequent implication that the organ is compound.  Becher long ago ('82) pointed out that the separation into
two parts under manipulation was an artifact.  Morphologically for
the imago of Culex (pl. 16, fig. 46–47) and for the imagoes of other
flies (Kräpelin, '82, '83) the labrum is a unitary structure.  During
the metamorphosis of Culex also, the labrum is formed as a simple
tube.  The canal (pc) on its ventral face is moulded by the infold-
ing of the ventral wall.  The labrum receives a single muscle which
probably serves for a retractor and depressor.  This muscle is inserted on the boss of chitin at the base of the labrum and its fibers
arise from the dorsal and posterior walls of the clypeus.  It may be
called the labral muscle (pl. 12, fig. 2).  This muscle has been described as "labral muscle" (Annett and Dutton, : 01), "elevator of
labrum-epipharynx" (Nuttall and Shipley, : 01–: 03), "retractor of labrum" (Giles, : 02), and "retractor partes productae" (Meinert, '81).
It corresponds in part to the "pharyngeal muscle" of Dimmock
('81).

The mandibles are wanting in the male mosquito and are slender,
delicate lancets in the female.  Dimmock ('81) has figured these
organs in the cross section of the mouthparts as lying beneath the
hypopharynx.  Giles (: 02) gives a similar arrangement in one figure while in another he shows the mandibles above the hypopharynx
at the sides of the labrum.  This last is the proper position for the
stylets when at rest.  The mandibles are morphologically dorsal to
the hypopharynx.  The other position is due to misplacement during the processes of sectioning.  Each mandible is retracted by a
mandibular muscle (mnd m) which arises on the "head" of the
tentorium and is inserted on the base of the stylet.  As these muscles approach their insertions short fibers connect them with the

adjacent walls of the head, but I am not certain whether the muscle is augmented by these fibers or instead partially inserts on the wall of the head.

The maxillae of the male mosquito are no stronger than the mandibles of the female, but the maxillae of the latter sex are powerful cutting lancets, armed near the apex with recurved teeth along the outer border of the blade. This border is thin and its chitinous cuticle is delicate. The inner border of the blade is broad and its chitin is thick (pl. 16, fig. 47). Proximally, the inner border terminates with an articular spur. From the point where the palp and the outer border of the maxilla blade unite, an apodeme enters the head and extends almost to the occipital foramen (pl. 12, fig. 2, *apo*). The maxillae have a complex musculature which must ensure a considerable freedom of movement, although it is possible to assign functions to the different muscles only in a general way (pl. 12, fig. 2; pl. 13). The most prominent muscle in sections lies parallel to and outside of the maxillary apodeme. It arises from the under side of the lower end of the tentorium and the adjacent walls of the head and is inserted on the base of the palp and apodeme. While the general action of the muscle must be the retraction of the maxilla, it would seem that it will also tend to divaricate maxilla and palpus. To adopt Dimmock's ('81) and Nuttall and Shipley's (: 01–: 03) name, this muscle is the retractor of the maxilla. Another important muscle arises from the "head" of the tentorium and is inserted on the articular spur of the maxilla. As it also sends fibers to the dorsal angles of the base of the labium, I call this muscle the double retractor. From its position it is probably the chief retractor of the maxilla blade, which cuts on the up stroke. It has an extensive tracheal supply.

Two pairs of muscles are found which probably serve to protract the maxillae. The larger of these, the protractors of the maxillae, are inserted on the free ends of the apodemes (*prot max*) and extend forward, inward, and upward to the tentoria. The origin is from the under surface of the strut just below the "head." The smaller pair of muscles, the maxillo-labials (*max-li*), are inserted towards the ends of the apodemes, and run forward along the inner side of the apodemes to the ventral border of the labium. In addition to these larger muscles, each maxilla receives on the articular spur a short band of fibers from the adjacent wall of the pharynx. Per-

haps these maxillary muscles counteract the divergent pull of the
retractors of the maxilla.   The retractors of the maxilla were
described by Annett and Dutton (: 01) as "muscle attaching maxillary process to occipital region of the skull."   Possibly the double
retractor is to be seen in Meinert's "retractor scalpelli" ('81, fig. 15).
In the same figure the "protractor scalpelli" may represent the protractor of the maxilla.   This protractor is described by Annett and
Dutton and figured in figure 1, plate 18, but is not named.   The
maxillo-labial muscle is described by Annett and Dutton as "muscle
to base of labium."   It is less readily traced in other studies and
may be either the "protractor scalpelli" or the "retractor scuti ven-
tralis" of Meinert ('81, fig. 16), while it is possible that the muscle
represented in figure 1, plate 8, of Nuttall and Shipley's account
under the name "protractor maxilla" represents it.   I have not
found any muscle that will answer to Meinert's "retractor scalpelli"
as shown in his figure 16.

The hypopharynx has a central "body," the mid-dorsal line of
which is traversed by the salivary gutter, and thin lateral "wings."
The hypopharynx is continuous with the labium at its first appearance in the pupa (Hurst, '90) and never becomes a separate stylet
in the male insect, so that the salivary gutter traverses the dorsal
face of the labium.   At the base of the hypopharynx, the salivary
gutter becomes continuous with the salivary pump, or as it is more
often termed, the "receptacle of the salivary duct."   This pump
(pl. 12, fig. 1) consists of a chitin-lined cup, open above and in front
into the salivary gutter, and closed behind with a plate of thin
chitin overlaid by a mass of cells.   This plate furnishes insertion
for two muscles and is pierced by the orifice of the salivary duct.
When the muscles attached to the plate contract, they draw the
plate back, out of the cup, and saliva can flow in from the salivary
duct.   On their relaxation, the elasticity of the chitin causes the
plate to re-enter the cup, forcing the contained saliva along the
salivary gutter into the wound.   Nuttall and Shipley (: 01–: 03) and
Annett and Dutton (: 01) have carefully described the similar pump
of Anopheles.   In *A. punctipennis* the pump appears to be of the
same size as that of Culex in specimens of similar length.   The
hypopharynx receives muscle fibers from the mandibular muscles.
The muscles which operate the salivary pump, the hypopharyngeal
muscles, will be described in connection with the pharynx.

The labium, the largest of the mouthparts, is cylindrical, with a deep furrow along the dorsal surface. It terminates in a median ligula and two lateral labellae. The labellae are moved by small muscles, labellar muscles (Annett and Dutton, :01), from the walls of the distal part of the labium itself. The organ as a whole is not well provided with muscles. Proximally, the dorsal angles where it meets the head, receive fibers from the double retractor muscles and on the ventral border the maxillo-labial muscles arise. It is probable that the return of this organ to position after the displacement during blood sucking, is accomplished without the activity of muscles. The lumen of the labium is traversed by the two labial tracheae and by two nerves (li n) derived from the infraesophageal ganglion of the brain.

The remaining muscles of the head fall into two groups : those associated with the various parts of the alimentary canal, and those supplying other organs. The former class will be discussed in connection with the parts of the alimentary canal which they supply. The latter group may be briefly noted. Each antenna receives scattered fibers from the frons and two large muscles from the "head" of the tentorium. One of these, the inner antennal muscle, is inserted on the inner angle of the base of the antenna. On contraction it must tilt the antenna forward and inward. The other muscle, the outer antennal muscle, is inserted on the outer posterior angle of the base of the antenna and hence on contraction must tilt the antenna backward and outward. This muscle is especially large in the male mosquito. On the outer face of either tentorium is inserted a slender muscle which enters the head through the occipital foramen. The muscles of this pair may be called the tentorial muscles and they correspond to a muscle labeled "retractor of the maxilla" by Nuttall and Shipley in their figure 23 on plate 8, and to the "salivary muscle" of Christophers (:01). The "retractor maxillae" in the remaining figures given by Nuttall and Shipley is the retractor of the maxilla of my series. The subocular muscles are a pair of minute muscles that arise on the floor of the epicranium nearly 0.1 mm. behind the bases of the tentoria and are inserted on the struts near the origin of the protractors of the maxillae. These muscles were not recognizable in the specimen from which figures on plate 13 were drawn, but have been inserted in two instances (figs. 10 and 11) from another series of sections.

The head of the Anopheles mosquito is relatively higher and broader in the epicranial region than the head of Culex and the pre-ocular region slopes downward. This makes the muscles run forward at different angles from those in Culex and sections of the two forms appear somewhat unlike, although, with a single exception, the same muscles are present in both. The muscles of the mouthparts and head, outside of those supplying the alimentary canal, differ from the muscles of Culex as follows. The retractors of the maxillae arise more laterally and fully 30 $\mu$ caudad from the bases of the tentoria, so that their origin is wholly from the walls of the head and not partly epicranial and partly tentorial. The posterior or apodeme ends of the protractors of the maxillae also, lie farther back and the upper or tentorial ends are in the same level as the double retractor muscle. Hence both protractor maxillae and double retractor muscles may be present in one section, which is not the case with Culex where the origin of the protractors is distinctly caudad from the origin of the double retractors (pl. 13, figs. 7, 15). The subocular muscles are much larger, and arising farther back in the head, they are less likely to be mistaken for fibers of the retractors of the maxillae. They are inserted on the tentoria immediately caudad of the origin of the double retractor muscles, near the insertion of a pair of minute transverse muscles, not represented in Culex, which arise on the walls of the epicranium dorsal to the origin of the anteriormost lateral dilator of the esophagus.

*The fore gut.* — The stomodaeal portion of the alimentary tract of insects is subject to great variations with respect to the number and character of the regions that are differentiated in it. It may well be considered doubtful if homologies between regions can legitimately be sought beyond the limits of any one order. Hence a nomenclature for the various parts need only be self-consistent within a single order and may properly be subject to individual preferences. The stomodaeum of the imagoes of the Diptera offers especial obstacles to the establishment of a system of names that shall meet all conditions, for there is a considerable range of variation within the group. For example, in Musca (pl. 14, fig. 16) the labral or proboscis canal (*pc*) is succeeded by an intracephalic subclypeal pump (*ph*) which shows distinct anterior and posterior regions. Then the esophagus (*oes*) commences and hence extends through the circumesophageal nerve-collar or "brain." In other flies with a similar succession of proboscis canal, subclypeal pump, and extensive esophagus, the differentiation of the subclypeal tube into two regions appears to be wanting. In contrast to the arrangements in these flies, the Tipulidae (pl. 14, fig. 18) have a chitinous "pump" in the region of the brain (*pump*) and the esophagus is restricted to the extreme rear of the head. The Culicidae (pl. 14, fig. 17) resemble this last mentioned family with the exception that the "pump" is dilated behind the nerve-collar instead of being of nearly uniform diameter throughout its length. The Simuliidae possess a "pump" which is not markedly dilated at any point, but it is distinctly divided into a preneural and a postneural portion. This subdivision foreshadows the arrangement found in the Asilidae and Tabanidae (pl. 14, fig. 19). Here the postneural part of the "pump" is vestigial while the preneural part is strongly developed. If a nomenclature is to be satisfactory for all cases it clearly must take into account three differentiated regions of the stomodaeum, in addition to the unspecialized esophagus: the proboscis canal of the labrum (*pc*), the subclypeal tube (*ph*), and the more posterior chitinized "pump." Of these the subclypeal tube and the pump may be subdivided. Two additional factors must be considered. The subclypeal tube appears to be an essentially homologous structure throughout the group of the flies and the pump probably corresponds to the anterior end of the elongate esophagus of those flies where only a subclypeal tube (*ph*) is differentiated, *e. g.*, Muscidae. A system of names which groups the subclypeal canal and pump under a single expression is therefore inadvisable, as for example Christophers' and Annett and Dutton's accounts where these regions collectively form the "pharynx" or Hurst's description in which the terms "buccal cavity" and "pharynx" are used indifferently for all the specialized intracephalic stomodaeum.

Two systems of nomenclature for the Dipteran stomodaeum which are free from the objections cited, exist. The older seems to have come into being by simple extension of the names given to the parts in Musca, *i. e.*, "pharynx" for the subclypeal tube, and "esophagus" for all the remaining fore gut. Thus Meinert ('81) calls the pump of Tabanus and Asilus "pars prior oesophagi" or "pars tumida oesophagi." With Culex or Tipula this region is simply "oesophagus." Dimmock ('81) with respect to the parts in Culex, the most complicated type of Dipteran alimentary canal that he studied, refers to the pump as "oesophageal bulb" and to the subclypeal canal as "pharynx." On the other hand, Nuttall and Shipley (:01–:03) in their careful study of Anopheles term the subclypeal canal "buccal cavity" and the pump "pharynx." The older of these nomenclatures has decided advantages over the newer system used by Nuttall and Shipley in that it recognizes the probable equivalence between the pump and the anterior end of the "esophagus" of such a fly as Musca, and in that it gives the expression pharynx — which is generally employed in insect morphology to indicate the first differentiated intracephalic region of the stomodaeum — to the constantly present subclypeal canal rather than to the inconstant pump. It also leaves the term buccal cavity vacant and available for any subregion of the subclypeal canal which it may seem best to delimit.

A study of the larvae of the Diptera in the above connection offers no assistance, but rather complicates the problem. A majority of Dipterous larvae have a fore gut that is practically undifferentiated throughout its cephalic extent. In other forms, well marked regions are present. But these are not always comparable, even in closely allied genera, e. g.. Culex and Corethra. Further, the larval regions may or may not coincide with the regions of the imaginal tract. The wriggler of Culex has a buccal cavity and a pharynx, the latter being imperfectly distinguished from the esophagus below and at the sides. But the larval buccal cavity and pharynx both go to form the subclypeal canal of the imago and the pump of the latter stage is derived from the anterior end of the esophagus of the larva. Such peculiarities find their explanation in the extremely adaptive character of the Dipterous larva, but this in turn makes it unwise to lay stress on larval relations, in an attempt to obtain a system of names for the imaginal parts.

Employing the nomenclature just chosen, the stomodaeal portion of the alimentary canal of the mosquito shows successively: the labral or proboscis canal, the subclypeal tube or pharynx, the pump or antlia, and an esophagus which extends through the occipital foramen into the thorax. The midgut is differentiated into cardia and stomach or mid-intestine. The hind gut is modified to form an ileo-colon and a rectum. Valves are developed at the union of pharynx and antlia, pharyngeo-esophageal valve, at the posterior end of the antlia, at the junction of esophagus and cardia, esophageal valve, and at the point where the stomach and ileo-colon come together.

According to Dimmock ('81) both labrum and hypopharynx take part in the formation of the proboscis canal, the latter stylet forming the floor. Kräpelin ('82) on the other hand, considers that this canal is wholly labral in origin. It was not possible to decide between these views. The shelf formed by the produced lower internal angles of the labrum (pl. 16, fig. 46–47) favors Kräpelin's view, but in this case the delicate character of these plates would seem to require that the hypopharynx should be drawn up against them during blood sucking and so indirectly complete the proboscis canal ventrally.

Since the anterior wall of the clypeus is forward of the bases of the mouthparts generally, the labrum meets the head some distance in front of the point where the mandibles, maxillae, hypopharynx, and labium are inserted. In this region the section of the proboscis canal is that of an arch, and the hypopharynx unquestionably supplies the flattened floor (pl. 13, fig. 3). Caudad of the bases of the mouthparts (pl. 13, fig. 4) the canal may be considered as intracephalic and as pharynx. At this point the thick chitin of the side walls of the canal thins abruptly and in material cleared with caustic the proboscis canal appears to terminate with a spur on either side. There is no real point of demarcation, however, between the proboscis canal and the pharynx as I have limited them, except the completion of the tube as the mouthparts join the head and the formation of these seeming spurs. The chitin of the floor of the canal is already thin and so harmonizes with the thinned walls. On the roof of the canal the heavy chitin extends for a considerable space into the head, thinning out gradually from the sides toward the midline, and so forming a tongue of heavier chitin amid the weaker chitin. Nuttall and Shipley's description of this structure, anterior hard palate in Anopheles as "somewhat in the form of a trowel" is equally vivid for Culex, since its arch is steeper than the arch of the pharynx roof generally (pl. 13, fig. 5). Just before this palate terminates, its surface is roughened by a few minute conical spines. The proboscis canal receives immediately anterior to the "spurs" at its proximal end, a pair of muscles from the alae of the clypeus. These are the "epipharyngeal muscles" (Annett and Dutton, :01) and the region of their insertion closely corresponds to that occupied in the larva by an epipharynx-like structure. Meinert ('81) figures for Culex muscles joining the alae of the clypeus to the roof of the box. I have not been able to find these in three species of Culex at my disposal, and they are not present in *Anopheles punctipennis*.

The pharynx extends from the union of the mouthparts with the head to the anterior end of the esophageal antlia. It is developed as a pumping organ and the character of its walls and the form of its cross section vary in different levels. At first, in the region of the anterior hard palate, the floor and walls have a thin intima throughout and this steadily involves more and more of the roof of the canal as the anterior hard palate narrows. At the same time the floor of the pharynx becomes curved, so that by the time the trowel-shaped hard palate has vanished, the pharynx consists of a curved ventro-lateral plate and a flattened dorsal plate, the two meeting in high dorso-lateral angles (pl. 13, fig. 6). The section of the tube is either crescentic or that of an inverted arch, according to the position of the dorsal plate as it curves down into the lumen or is drawn upward by muscular action. For a space after the hard palate vanishes, both walls and roof remain thin. Then the dorsal plate becomes uniformly thicker. Finally, thinning again, it forms the pharyngeo-esophageal valve at the entrance to the antlia. Hence the dorsal plate or roof of the pharynx is differentiated into four regions: an area of thin chitin with a median tongue of thicker chitin, anterior hard palate; a region with uniformly thin chitinous intima, soft palate; a region with uniformly denser intima, posterior hard palate; and a narrow area of thin chitin forming a valve. The valve and posterior hard palate regions are derived from the "pharynx" of the larva. The remainder of the pharynx and the proximal part of the proboscis canal are formed from the "buccal cavity" of this earlier stage.

The chitinous intima of the floor and side wall of the pharynx is uniformly thin except in the region of the posterior hard palate. Here, beginning first along the border of the dorso-lateral angles, an area of thicker chitin forms on either hand and widens to involve all the wall of the canal for a space. Then narrowing again, these areas terminate with stout spurs which project past the anterior end of the antlia. At the point where they are most extensive, they meet across the floor of the pharynx and in the female insect a delicate crest projects downward from this traverse. The male Culex has the same arrangement of thickened areas but lacks the crest. The development of these thickened areas in the walls of the pharynx near its union with the antlia must make the ventro-lateral plate rigid at this point. The antlia has also at its anterior end rather unyielding walls. Hence the thin dorsal plate beyond the posterior hard palate forms an efficient pharyngeo-esophageal valve, dipping into the lumen of the canal through elasticity and being withdrawn by muscles.

At its anterior end the pharynx for a space is firmly bound to the walls of the head by an ascending chitinous plate on either side. These plates are continuous with the cuticle of the head and the intima lining the pharynx, and represent intracephalic continuations of the plates formed by the union of the genae with the clypeus (pl. 13, fig. 3–5). The plates soon degenerate into low crests along the dorso-lateral angles of the pharynx and fade out.

At its hinder end the pharynx is held in place by two pairs of muscles. The shorter of these, the lateral pharyngeal muscles, run outward from the external face of the spurs at the rear of the pharynx to the tentoria. The others, the ascending pharyngeal muscles, pass from the internal face of the spurs to the vertex of the head.

The pharynx as a pumping mechanism is constructed on the same principle as the salivary pump or antlia. The lumen is diminished by the inspringing of the chitinous dorsal plate, which is withdrawn again by the action of muscles. The muscles which lift the roof of the pharynx in this pumping action may be called the elevators of the palate (Nuttall and Shipley, :01–:03). When compared with the homologous muscles of many flies these muscles appear relatively weak. They are considerably specialized, however, as there are five distinct pairs, instead of a single mass or at most a larger anterior and a smaller posterior section. When this latter differentiation occurs the posterior section is called the protractor of the pharynx (Meinert, '81). Possibly it finds its equivalent in the fifth pair of the Culicid muscles. Of the five pairs, the anteriormost insert on the rear of the anterior hard palate region, the next on the soft palate, the three posterior pairs on the hard palate. The pharyngeo-esophageal valve is elevated by a pair of valvular muscles from the frons. These lie caudad from the buccal ganglion. The valvular, ascending pharyngeal, and lateral pharyngeal muscles are described by Nuttall and Shipley, but are not named. The elevators of the palate are the "musculi antliae pharyngis" (Meinert, '81), the "pharyngeal muscles" (Annett and Dutton, : 01; Dimmock, '81, in part), and the "protrusor muscles of the labrum" (Giles). The last identification is based on the figures. The accompanying description is obscure.

The pharynx serves as origin for two pairs of muscles. Near its anterior end the maxillary muscles arise and near the posterior end the hypopharyngeal muscles. These latter operate the salivary pump. With the female of Culex they have an extensive origin from the walls of the pharynx between the ventrally projecting crest already described and the neighborhood of the third pair of elevators of the palate (pl. 12, fig. 1, *hyp m*). They arise in part also from the lateral borders of the crest, but not from its median surface. In the male the smaller muscles have a limited origin in the level of the third pair of elevators of the palate. No muscle with the position assigned by Giles to his "muscle opening the salivary valve" could be found in either Culex or Anopheles and the account of the salivary gland given by him is misleading in other respects. The hypopharyngeal muscles correspond to Giles' "muscles closing the salivary valve," to Meinert's "retractores receptaculi," and to Annett and Dutton's "muscle to the salivary receptacle." Nuttall and Shipley describe these muscles but do not name them.

The structure of the pharynx described for Culex is closely paralleled by that found in the malarial mosquito. There is the same succession of anterior hard, soft, and posterior hard palates and valve, but for *Anopheles maculipennis* as described by Nuttall and Shipley, the structures on the intima in the region of the tip of the anterior hard palate are more complicated than they are in Culex. Annett and Dutton describe for *A. costalis* a complex arrangement of hair-like processes on the ventral plate of the pharynx below the pharyngeo-esophageal valve. Nuttall and Shipley record that they failed to find such structures in the species of Anopheles that they studied, *A. maculipennis*, and I do not find them in *A. punctipennis*. The bracing of the pharynx to the walls of the head anteriorly and the formation of heavily chitinized areas and terminal spurs posteriorly are similar in both genera and in Anopheles also the ventral crest is developed as a partial origin for the hypopharyngeal mus-

cles. It would appear that the pharynx of Anopheles has somewhat more sharply marked angles than the pharynx of Culex and
may therefore be a more effectual pump. The muscles are the
same as in Culex: hypopharyngeals from the walls and crest of the
pharynx, five pairs of elevators of the palate with the same relations
to the soft and posterior hard palates, epipharyngeals from the alae
of the clypeus, lateral pharyngeals, ascending pharyngeals, and
valvulars. The ascending pharyngeals arise from the median internal crest alone and not from crest and vertex of the head as is the
case with Culex.

The antlia of the female mosquito is pyriform with the bulb
behind the nerve-collar. The chitinous intima which lines this part
of the alimentary canal is so modified that it is convenient to speak
as if it were composed of three racquet-shaped plates, one dorsal,
the others ventro-lateral in position.

The borders of the plates are reflected and a thin band of chitin
is included between each adjacent pair of plates. The plates can
spring inward until they almost obliterate the lumen of the canal
(pl. 13, fig. 12) or be drawn back by muscles until the lumen of
the bulb has a nearly circular section. As the plates terminate
posteriorly they are clothed by a dense coat of fine bristles and
this development continues for 10 $\mu$ or more on the thin intima of
the esophagus. The postneural bulb is the active part of the pump.
Here the dorsal plate receives two large dorsal dilator muscles from
the vertex of the head. These muscles are compressed and in an
average female of *Culex stimulans* their greatest diameter was 60 $\mu$,
the lesser diameter 31 $\mu$. In a male of the same size the corresponding diameters were 60 $\mu$ and 15 $\mu$. These muscles correspond
to the " musculi superiores antliae " (Meinert, '81) and to the " supe-
rior pumping muscles " (Christophers, :01), " post dorsal dilators "
(Nuttall and Shipley, :01–:03). Either lateral plate receives five
lateral dilator muscles from the sides of the head. These muscles
trend forward and upward to their insertions and are compressed.
In an average female of *C. stimulans* the greatest diameter of the
first to the most posterior muscle was 54, 56, 72, 54, and 44 $\mu$
respectively and the lesser diameter 27, 20, 37, 18, and 20 $\mu$. The
total cross section of the female's muscles is nearly twice that of the
muscles of the male. The lateral dilators have been described as
" musculi inferiores antliae " (Meinert, '81), " lateral pumping mus-

cles " (Christophers, : 01), "muscles of the pumping organ " (Annett and Dutton, : 01), "dilator muscles " (Giles, : 02): Dimmock figures and refers to the lateral and dorsal dilator muscles but does not name them.  In the anterior, preneural, part of the pump the ventral uniting strip between the ventro-lateral plates — always broader than the two dorso-lateral strips — becomes wide and flat, so that the antlia in this region is four-sided (pl. 13, fig. 8).  This must diminish the movement of the walls and dorsal plate.  At this point the dorsal plate receives a vestigial dorsal dilator muscle (*ant dil*).  In Anopheles, where the activity of this part of the pump is less diminished, two dilator muscles reach the dorsal plate, "anterior dorsal dilators " (Nuttall and Shipley, : 01–:03).

The epithelium of the proboscis canal, pharynx, and antlia is inconspicuous.  It is flattened over the proboscis canal and antlia more than over the pharynx.  The intrinsic muscle coats are not evident, except over the hinder end of the antlia where a powerful sphincter muscle is developed (pl. 12, fig. 2, *sph m*).

The male of Culex has a more attenuated antlia with a relatively smaller bulbous portion than the female.  The total capacity is less than that of the female's antlia even when the latter insect is represented by a smaller specimen.  Computed by Cote's rule, the largest male antlia found in *C. stimulans* had a capacity of approximately 0.0008 cu. mm.  The smallest female antlia of the same species measured 0.002 cu. mm., and the largest 0.004 cu. mm.  The time taken by a mosquito of this species to draw up its meal of blood depends on the nature of the spot pierced, averaging according to my observations about one minute for a light and nearly half as long again for a heavy meal.   But under especially favorable conditions the lesser meal can be taken in 50 seconds, the greater in 65 seconds.  Since a heavy meal consists of nearly 4.5 cu. mm. of blood, a very rapid action of the antlia would seem to be required, especially if this pump empties itself at each stroke.  More probably, however, it acts to maintain a continuous rather than an intermittent flow.

The antlia of Anopheles corresponds closely to that of Culex in structure.  The most important difference is the greater functional activity of the anterior or preneural part as a pumping organ with an accompanying development of the anterior dorsal dilator muscles.  As a result, the antlia of Anopheles is more uniformly cylindrical than the antlia of Culex. This point of difference is of interest in connection with the varied specialization of the antliae in the Diptera. On a basis of their pumping mechanisms the group can be separated into two main groups, which may be termed monantlial and diantlial. The former division includes the Muscidae (pl. 14, fig. 16), the Syrphidae, and other families the members of which possess only the pharyngeal pump. The diantlial group have both a pharyngeal pump and an antlia. Among these flies several forms of antlia are found. For example, Culex (pl. 14, fig. 17) and Anopheles have an antlia which is best developed behind the "brain," i. e., it may be described as postneural in type. As already noted, the pump of Anopheles is less markedly postneural than that of Culex, approaching what may be called the amphineural type, where it is nearly cylindrical throughout. The Tipulidae are perhaps the best example of this type of antlia (pl. 14, fig. 18, *pump*). Simulium, while amphineural with respect to the form of its antlia, has this region differentiated into distinct pre- and postneural sections. The antlia of the Tabanidae (pl. 14, fig. 19) and the Asilidae is not only subdivided as in Simulium, but the preneural part alone is functional, the post-neural section being rudimentary. This type may be called preneural.

The tracheation of the head is interesting with respect to the supply to certain parts of the alimentary canal and muscles. Thus the extensive tracheal supply to the elevators of the palate shows that these muscles are of considerable use, and that the pharynx, although overshadowed by the great esophageal pump, is yet by no means unimportant as a suctorial bulb. The thoracic portion of the alimentary canal is paralleled on either side by a tracheal trunk. From either trunk two tracheae are given off into the head, the external and internal tracheae (pl. 13, fig. 12). These lie close to each other until they have passed the dilator muscles and then part company. The external trachea ascends into the cleft between the optic ganglion and the "brain" (pl. 13, fig. 10), supplies these nervous masses and the protractor muscle of the maxilla, and then breaks up into branches some of which run in the mesial plane of the clypeus to supply the labral muscle and the elevators of the palate. The internal trachea follows the ascending line of the tentorium and its terminal twigs reach the head of the strut. But before this point is attained, the trachea has spent itself in two large branches. The first to arise, the labial trachea (*li tra*), runs forward between the double retractor and retractor of the maxilla muscles and enters the labium. It gives branches to the double retractor, the retractor of the maxilla, the salivary pump, and probably to the mouthparts generally. One prominent tracheole follows the double retractor muscle, so that this muscle has an extensive tracheal supply. The second branch of the internal trachea (pl. 13, fig. 6-9) trends sharply upward and divides to supply on the one hand the maxillary palpus and on the other the lateral pharyngeal and antennal muscles and the elevators of the palate, which thus receive a double tracheal supply. A median tracheole reaches the labral muscle at its origin.

The esophagus is mainly thoracic in the female Culex and almost wholly so in the male, owing to the more truncate head which leaves less space beyond the antlia. The esophagus is a slender canal with thin walls, lined with a delicate intima which is smooth except for the area of bristles at the anterior end. The walls are composed of a single layer of flattened epithelial cells, the boundaries of which can hardly be distinguished in sections. The usual muscle coats are present, but poorly developed. Near the posterior end of the esophagus three diverticula arise, two from the dorso-lateral and one from the ventral wall of the gut (pl. 14, fig. 20, *f res*). Then the esophagus terminates with the esophageal valve, a shallow curtain which dips into the first region of the midgut, the cardia. Within this valve the otherwise insignificant circularis muscle coat is specialized to form an annular muscle. The valve appears to be without an external blood sinus and its walls have an epithelium similar to that of the esophagus. Unless the cardia is greatly distended, the point where the epithelium of the esophagus ceases is very noticeable at the shoulder of the cardia near the termination of the reflected face of the valve (pl. 16, fig. 41).

The dorsal esophageal diverticula are small and lie close to the dorso-lateral walls of the prothorax. The ventral pouch is large and may extend into the abdomen. It does not appear to be as large relatively as the corresponding pouch in Anopheles. Usually all three sacs are filled with bubbles of air, and I have never found anything else in them. Nuttall and Shipley (:01-:03), however, have made a careful study of the possible functions of these pouches in Culex and seem to have conclusively demonstrated that they serve as food reservoirs. Other investigators record the detection of blood in the sacs of the mosquito and the similar pouches developed in other flies occasionally contain food. In view of these facts there seems to be no good reason for denying that these diverticula serve as storage places for food. The name "food reservoirs" (Nuttall and Shipley, : 01–: 03) or "esophageal diverticula" (Granpré and Charmoy, : 00) should replace the older and misleading terms "aspiratory vesicles," "suctorial vesicles," etc.

When one of these diverticula is extracted and is examined under a microscope, the only structure noted at first is an irregular and varying mesh of delicate striae. More careful study makes it clear that these "striae" are wrinkles or folds of the delicate intima that lines the pouch, varying in size and position with the torsion or tension of the particular part under observation. By staining the fresh tissue with Bismark brown, distinct muscle fibers become visible and the nuclei of the epithelium appear (pl. 16, fig. 42). With a $\frac{1}{12}$ oil-immersion lens the epithelium itself can be distinguished in places as a granular cloud. Whether the muscle bands have as regular a distribution with Culex as the similar bands described by Nuttall and Shipley on the ventral sac of Anopheles, was not determined. The "striae" sometimes pass across a muscle without alteration, but usually, of course, they are more numerous along the line of the muscle, interlacing and radiating in groups from its borders. Under a low-power lens this gives them a strong resemblance to muscular fibrillae. The appearance vanishes when a higher magnification is used, but should be noted, since, with some reagents — e. g., methylin blue — the color fills the "striae," seemingly staining them. If a pouch is torn the excessively wrinkled intima gives a fibrous appearance to the fragments. This is probably the basis for Giles' assertion that these organs are composed of finely branched tracheae. "The fibers are neither more or less than extremely elastic and distensible tracheae, which swell out into bubble-containing dilatations. .... Apart from a few loose connective elements the sacs consist of nothing else than these curiously modified tracheae." In connection with this statement with respect to structures that are obvious diverticula from the esophagus and have a less extensive tracheal supply than any other part of the alimentary canal, except possibly the pharynx and antlia, Giles also holds that there are two ventral sacs, that there is no organic connection between the esophagus and the sacs, and that the pouches have an aerostatic function. Nuttall and Shipley (: 01–: 03) have thoroughly exposed the inaccuracy of this account of the sacs of Anopheles. It would be equally inapplicable to Culex. The two genera show no essential differences with respect to the structure of the food reservoirs.

*Salivary glands.*— The epithelium of the salivary duct is thin but columnar and the duct has a trachea-like aspect owing to annular thickenings of its intima. It forks within the head and in the thorax each branch supplies a tripartite gland. The duct within each of the three acini of the gland is (*Culex stimulans*) of uniform diameter, has a delicate smooth intima, and is traceable to the apex of the acinus where it ends blindly. The acini are elongate and as their first describer Macloskie ('88) and later writers have noted, the central and lateral divisions are unlike in appearance. Macloskie gives no evidence to support his suggestion that the central acinus secretes the poisonous element of the saliva. This acinus (pl. 14, fig. 23, *ca*) is slender, cylindrical, and its nuclei are prominent, surrounded by granular cell contents. In sections the larger part of each cell is filled with secretion which, as precipitated by the reagents, stains lightly with haematoxylin and is coarsely granular. The nuclei are pushed over against the basement membrane. The lateral acini apically are of the same diameter as the central acinus, but their bases are distended. Their cells are filled with a transparent secretion of high refractive index. Toward the base of the gland nearly every cell contains a huge ovoid vacuole filled with the secretion which forces the nucleus against the basement membrane. Sections show clear spaces where these vacuoles were situated. Nearer the apices of the acini the vacuoles are smaller and less numerous.

The differences in the appearance of the lateral acini and the central acinus are due to the different substances secreted and stored in the cells. During the later pupa stage the distinctions are wanting and all parts of the gland resemble the central acinus of the imago as viewed by transmitted light. Sections, however, show no secretion in the cells. The differences make their appearance a short time before the imago emerges.

*The midgut.*— The part of the midgut which receives the eso-phageal valve (pl. 14, fig. 20) is best called the cardia. As with most other flies, the region is not definitely marked off from the

stomach in either Culex or Anopheles.   Sections show only a gradually increasing affinity for haematoxylin on the part of the cells as
the esophageal valve is approached.   In shape and other characters
the cells resemble those of the stomach generally and there is
no alteration in the muscle coats.   The stomach extends into the
seventh abdominal segment and is the longest single region of the
alimentary canal.   Its thoracic end is a narrow tube, but the abdominal part is wider and capable of great distension.   Here is stored
the meal of blood while it is being digested.   The wall of the
stomach has an epithelium composed of small cubical cells with oval
nuclei and granular cytoplasm (pl. 14, fig. 21, *st*).   The free border
of the cells is more transparent than the remaining area.   Probably
this corresponds to the "striated border" described by Christophers
(:01) for Anopheles.   The margins of the cells make a very sharp
line in sections, but I did not succeed in demonstrating a cuticle.
In the freshly removed stomach the nuclei are not visible and the
cytoplasm appears filled with refringent granules.   Over the surface
delicate circular and longitudinal muscles pass in an irregular net-
work which cuts the surface into more or less rectangular areas.
The epithelium often protrudes between the muscles forming
"islands" of one or more cells.   The stomach narrows suddenly
at the posterior end and a low valve is formed.   The tracheal supply is extensive and the interlacing vessels serve to retain the canal
in position.

The physiological dissimilarity between the stomachs of Anopheles and Culex in relation to the malarial parasite may possibly
be accompanied by visible histological differences, although none
have so far been described.   Conclusive proof awaits the elaborate
process of removing stomachs from mosquitoes of both genera when
in the same physiological state, and passing them through all the
reagents simultaneously.   I have not been able to obtain a sufficient number of Anopheles to do this.

*The hind gut.*— The anterior end of the first region of the hind
gut, the ileo-colon, is modified to form a low valve adjacent to
the valve (pl. 14, fig. 21) which terminates the midgut.   The Malpighian tubules open into the circular cleft between these two annular valves.   There are five Malpighian tubules, an unusual number
among the flies.   They extend forward for a space parallel to the
stomach and then recurve, their tips reaching into the last joint of

the abdomen.   In fresh material under transmitted light the cytoplasm is steel gray in color, the nuclei are transparent, and the cells
are filled with refringent granules.   Sections show a definite structureless border along the free edges of the cells.   This stains deeply
with " orange G " but I am not certain whether it is to be regarded
as a cuticle.

The ileo-colon is a canal of small diameter and its walls are thin.
The cytoplasm of the epithelial cells as seen in sections is often
vacuolated, especially toward the base of the cells.   A thin, distinct, cuticular lining is visible, and for a short space immediately
behind the entrance of the Malpighian tubules this is roughened
by bristle-like chitinous papillae which point caudad (pl. 14, fig. 21,
*ilc*).   This hirsute belt cannot be regarded as characterizing a
"region" in the gut of the imago and the area does not correspond
to any distinct region in the hind gut of the larva.   The longitudinal and circular muscles of the ileo-colon are fairly well developed
and like the stomach this part of the alimentary canal has an extensive tracheal supply.

Posteriorly, the ileo-colon passes imperceptibly into the rectum.
The depth of the epithelial cells of the rectum is variable and the
muscle coats are less developed than in the ileo-colon.   There are
six rectal papillae, each consisting of a cone of large epithelial cells
arranged about a common axis.   The tracheal supply to the rectum
is generous, especially to the papillae.   Tracheae derived from the
main longitudinal trunks that follow the alimentary canal on either
side run to each papilla, emit branches that course over the base of
the cone and adjacent walls of the rectum, and then one or more
branches enter the axis of the papilla.   These axial tracheae ascend
to the apex of the cone and divide, their branches curving back in a
somewhat tortuous course among the component cells.

## The Larva.

The larger and broader head and the elongate respiratory siphon
on the dorsum of the eighth joint of the abdomen are perhaps the
most obvious of the external differences that distinguish the wrigglers of Culex from those of Anopheles.   Otherwise the ground plan
is similar: the three thoracic segments are consolidated into a single

mass and there are nine abdominal joints, of which the eighth bears the respiratory siphon and the ninth four anal valves and a ventral keel of setae. Under the scope of this paper the head alone requires detailed description.

The head of the Culex wriggler (pl. 15, fig. 29-32) is flattened in the dorso-ventral plane. Forward of the antennae it narrows. Behind them its cross section would resemble that of a watch, a flattened oval. The dorsal surface is the more convex, descending rapidly in the rear, more gently at the sides, and with a comparatively steep slope forward from the antennae. The anterior margin of the head is bounded by a border line of thickened chitin (*bord l*) beyond which a shelf-like fold projects. The shelf bears two simple setae. The investing cuticle of the top of the head is strong and is reinforced by two thicker scythe-shaped areas (*sca*) which extend backward from the antennae. At pupation the cuticle splits along the inner margin of either area to the occipital foramen. On the ventral side of the head the transverse line (*tr l*) separates the pre- from the post-antennal region. This boundary consists of a narrow-linear band of dark, thickened chitin laterally and a low transverse fold mesially. Heavy cuticle extends over the post-antennal ventral region of the head and is traversed by two narrow-linear "lines" of dark, thickened chitin which pass from the transverse line to the border of the occipital foramen. These divergent lines support the ventral ends of the tentoria, the black, triangular, mental sclerite, and furnish a point of origin for the depressors of the antenna (*ant m*) and other muscles. At pupation the ventral head cuticle splits in the midline from the occipital foramen to the transverse line.

The relations of the pre-antennal region of the head are hard to describe. As noted, the dorsal surface is convex and heavily chitinized. The ventral face slopes sharply inward and downward toward the entrance to the pharynx, a point almost in the center of the head. Thus the pre-antennal region is in reality a broad rostrum which overhangs the mouthparts, situated near the transverse line below, slightly forward of the entrance of the pharynx (pl. 14, fig. 26). Its cross section would be the segment of a circle, the arc being represented by the convex dorsal surface. In the midventral line, about halfway between the anterior shelf-like fold and the entrance of the pharynx, a crest is formed. This bears four stout setae and is flanked by setose areas. Here the thin chitin characteristic of the ventral face of the rostrum generally, is strengthened by a triangular sclerite. I regard this structure as an epipharynx. It receives a pair of slender muscles (Raschke, '87) which arise on the top of the head and probably function as retractors.

The border line and shelf-like fold already described, mark the boundary between the heavier chitin of the dorsal and the thinner chitin of the ventral face of the rostrum, above and in front. But on either side, near the antennae, the thicker chitin involves part of the ventral face of the rostrum. These inflected areas of heavy chitin I call the black-spot areas, because each bears a conspicuous patch of pigment (*blks*). From either area a narrow-linear "line" of heavier chitin traverses the ventral face of the rostrum to strengthen the sclerite of the epipharynx. In the bay of thin cuticle bounded in front by the anterior shelf and border line, and posteriorly by the black-spot areas, lie the small, median palatum (*pal*) and the larger, lateral flabellae (*fl*). These are protuberances, densely clothed with fine hairs, the flabellae having in addition a peculiar arrangement of long yellow setae. Black-pigmented apodemes (pl. 15, fig. 81) that are continuous externally with the black-spot areas enter each flabella and the two flabellae are united by a transverse rod. To the apodemes the flabellal muscles (*retr fl*) are attached, two for each flabella, an inner and an outer, both retractor in function and acting simultaneously. When these muscles contract the seta-bearing area of the flabella is depressed and the setae come together in a brush whose tip points caudad. On their relaxation the flabellae protrude beyond the front of the head and the setae stand out in great yellow fans. The median prominence or palatum passively follows the movements of the flabellae, which are strictly synchronous, as far as I have observed.

*The mouthparts.*— The mouthparts (pl. 15, fig. 29, 32) of a Culex wriggler consist of huge maxillae, moderate sized, toothed mandibles and sclerites which represent labium and hypopharynx. The maxillae conceal the greater part of the under surface of the rostrum. The palp is minute. These appendages are not joined to the head by definite articulations as are the mandibles, and do not move as freely. Within the epipharynx above, mandibles, maxillae at the sides, and transverse line below, lies a funnel-shaped cavity leading to the pharynx. Into this space the tips of the flabellal setae dip when those organs are depressed. It may be called the buccal cavity, and within the transverse line four structures are successively differentiated from its floor: a delicate, hair-fringed crest, a black, triangular mental sclerite (*tr*), a complicated saddle-shaped sclerite with a median crest and lateral toothed spurs (*li*), and confluent with this last, an arch of heavy chitin which is pierced by the orifice of the salivary duct (*hyp*). The mental sclerite is braced to the divergent lines at its lower angles. The arch sclerite, and indirectly through it the saddle-shaped sclerite, are strengthened by a narrow-linear "line" of heavy chitin (pl. 15, fig. 31, *htr*) which crosses the thin chitin of the buccal cavity from the black-spot area of either side. These hypopharyngeal traverses give support also to the walls of the pharynx by short apodemes.

The mental sclerite was called "under lip" by Meinert ('86) and corresponds to the structures figured by Miall and Hammond for larvae of Chironomus, ('92, : 00), Miall ('93) for Dicranota, and Miall and Walker ('95) for Pericoma as "submentum." Raschke identified the fold that I have termed the transverse line with the mentum, "kinn." For Miall, the sclerites which correspond to the saddle-shaped sclerite of Culex are "mentum." With Culex, however, this sclerite sheathes the bud which forms the labium of the adult and is to be regarded as the larval labium. The mental sclerite would seem to be either a mentum or a submentum, as its hypodermis during the metamorphosis passes into the floor of the head (pl. 15, figs. 34, 36, *fold*).

The mandibles have a powerful musculature. Each receives a converger muscle on the lower internal angle and a divaricator on the lower external angle. These muscles arise by triple heads on the walls of the epicranium. Each maxilla has two muscles. One consists of two parallel bands and arises near the origins of the mandibular muscles (*dep max*), the other is single and arises from the divergent line (*retr max*). Both insert at the middle of the base of the maxilla. The action of these muscles was determined with difficulty, but by watching slow contractions in a dying larva it was found that the double band pulled the maxilla caudad and a little outward (ventrad) while the single muscle pulled the maxilla caudad and inward (dorso-mesad). Hence the double muscle may well be called the depressor of the maxilla, the single muscle the retractor of the maxilla. The labium is supplied by a pair of muscles running from the divergent lines to the hinder angles of the sclerite. They are depressor in action. Under the hypopharyngeal sclerite are a few scattered muscle fibers.

*The fore gut.*— The fore gut of the Culex larva is divided into buccal cavity, pharynx, and esophagus. The buccal cavity has already been described. Its epithelium is columnar and behind the epipharynx above and the hypopharynx below there is a considerable area of undifferentiated intima (pl. 15, fig. 32, *bc*). The pharynx is sharply marked off from the buccal cavity by a fine line of denser chitin and its flattened epithelium serves as a further distinction. The pharyngeo-buccal and pharyngeo-esophageal openings are not in line, since the esophagus leaves the pharynx rather from the floor, and in longitudinal sections the rear of the pharynx often overhangs the esophagus like a blind pouch (pl. 15, fig. 32). This appearance is an artifact, due to the contraction of the walls of the esophagus when the animal is killed. Actually, pharynx and esophagus are not sharply delimited. The general appearance when dissected out of the head, is as if the esophagus expanded to form a cup with high-arched sides (pl. 15, fig. 35) open widely in front where the walls pass to the buccal cavity and closed above by a dorsal plate. The arching borders where walls and roof meet may be termed the crests of the pharynx. They are approximated at the extreme rear of the pharynx (pl. 15, figs. 30, 33) and as they sweep forward in the high curve and descend again toward the buccal cavity, they diverge. The dorsal plate or roof of the pharynx therefore is narrowed posteriorly. When at rest the angle between the posterior ends of the crests measures about 60 degrees. But the side walls of the pharynx are mobile and may approach or spread apart from one another. The angle between them consequently varies, and the dorsal plate must accommodate itself to changes in the width of the space between the crests. This is accomplished by folding the plate inward along its median line. When the pharynx is dilated, the dorsal plate is almost plane, but when the walls of the pharynx approach each other strongly, the plate folds in until it nearly touches the floor beneath. The folding is most marked posteriorly and becomes progressively weaker toward the anterior end of the pharynx, where a transverse crescentic sclerite is developed in the dorsal plate and very possibly serves as a check (*cres*).

The pharynx has an elaborate musculature. The two halves of the dorsal plate are united by transverse or oblique bands (*dor ph*) which attach to a longitudinal sclerite developed on either side of the median fold. A pair of diagonal muscles (pl. 15, figs. 30, 33, *diag*) reach the pharyngeal crests near their posterior ends, from origins on the roof of the head, in each case on the side opposite to that of the insertion. At the summit of their curve the crests receive lateral muscles (*lat m*) from the scythe-shaped areas on the walls of the head, while nearer the border of the buccal cavity, several lesser lateral muscles (*less lat*) reach the dorso-lateral walls. At their approximated posterior ends the crests are bound to the epicranium by stout muscles, the retractors of the pharynx (pl. 15, figs. 30, 32). The hinder parts of the dorsal plate furnish insertion for two elevators of the dorsal plate (*el d p*), which descend from the epicranium. The lateral and ventral walls of the pharynx and the walls of the esophagus generally possess a well developed circularis muscle-coat and one of the most characteristic of the pharyngeal muscles is probably only a specialization of the circularis. This muscle girdles the pharynx from the angle where the crests meet, along under either crest and beneath the floor below. It may be called the girdle-muscle or cingulum (pl. 15, figs. 32, 35, *cin*). In front of the point where the cingulum passes under the floor of the pharynx, two slender muscles arise and extend back to the walls of the epicranium near the occipital foramen. These are the ventral retractors of the pharynx. They run in company with a number of slender muscles which emerge at intervals from beneath the circularis of both pharynx and esophagus. The complex that is formed by these muscles may be called the lateral dilator of the esophagus (*lat dil oes*). There are also two pairs of dilator muscles that insert on the dorsal wall of the esophagus (pl. 15, fig. 32, *ant dil, dor dil*). The larger of these lie above the supra-esophageal ganglion and closely parallel the retractors of the pharynx; the smaller lie beneath the ganglion.

When a wriggler is feeding the significance of the peculiar pharynx and musculature becomes apparent. It will be noted that the pharyngeal crests rapidly approach and separate so that the dorsal plate must be constantly infolded and withdrawn. The great fans of flabellal setae remain spread, but sweep downward in a small arc with a flickering motion. The mandibles and maxillae are also in incessant rapid motion. At intervals the flabellae are depressed and presumably any food that may be entangled on their setae is brushed off by the maxillae. When the pharynx is filled with food, an extreme contraction occurs, the crests approach very close to one another, and the food is forced into the esophagus. The retractor muscles of the flabellae do not produce the flickering movements just described, being concerned only with depression of the flabellae. They remain tense during the gentler movements. The retractors of the pharynx and the elevators of the dorsal plate are also passive. The motion appears to be caused by a slight to and fro movement of the entire pharyngeal and buccal regions, very possibly due to activity of the epipharyngeal and lesser lateral muscles and the lateral dilators of the esophagus. The changes in the pharynx itself, as shown by the movement of the crests, can be readily interpreted. I believe that the retractors of the pharynx and the elevators of the dorsal plate remain tense. This would make the elevation of the dorsal plate largely the passive result of the divergence of the crests as the pharynx walls separate. Probably the cingulum is the most important agent in approximating the walls of the pharynx. The diagonal muscles may assist and from their crossed position should be able to exert considerable leverage. Perhaps they are the chief factor in producing the extreme compression observed when the animal swallows, for any very marked contraction of the cingulum apparently would tend to close the pharyngeo-esophageal opening.

The head of the Anopheles larva is proportionally much smaller than that of Culex and, as Nuttall and Shipley (:01–:03) have pointed out, is rounded. "In fact the diameter from above downwards is very little less than from side to side, except anteriorly, where the dorsal surface slopes downward and forward." This rostral region is narrow and compressed at the sides, and the flabellae and palatum are carried well out beyond the apices of the mandibles and maxillae. The flabellae are relatively much smaller than in Culex. Hence they recall the flabellae of the larva of Simulium, but are not stalked as are the flabellae of this genus. The maxillary palpi of the Anopheles larva are very large and conspicuous. Between the maxillae the homologue of the mental sclerite is recognizable, and in sections labial and hypopharyngeal sclerites are found.

The elongation of the rostrum and the greater rigidity of its ventral surface require a different method of feeding from that employed by the Culex larva. Nuttáll and Shipley state that the maxillae, mandibles, and maxillary palp form the sides and floor of a cavity and that the flabellae "are suddenly bent back into this space, the mandibles and maxillae moving forward to meet them and at the same time opening out; they are then as suddenly released and fly back to their original position. This movement.... is repeated with great rapidity, often some 180 times to the minute producing a current sweeping in convergent curves towards the above mentioned cavity. From time to time the mandibles are approximated and the stiff curved hairs of their upper edge are run through the brushes" (flabellae). That this may aid in removing food from the flabellae or "brushes" is shown by the further observation that at intervals "the brushes disappear far into the mouth and are then slowly withdrawn, passing through the fine carding bristles on the inner face and anterior edge of the maxillae." In Culex, according to my observations, the food is borne into the space formed by the mandibles and maxillae by a flickering motion of the expanded flabellas and retraction of these organs is infrequent. The carding of the flabellal setae is irregular in occurrence and cannot be observed in detail owing to the unfavorable position assumed by the feeding larva. This subordinate *rôle* for the flabellae is the main difference between the feeding habit of Culex and that of Anopheles. Sections indicate that Anopheles has a pharynx of the same general form and musculature as Culex, but of relatively inferior size.

The esophagus (pl. 16, fig. 37, *oes*) has already been partly described. Its epithelium is flattened, especially toward the posterior end of the region. The chitinous intima is strong. Beneath the well developed circularis muscle-coat lies a longitudinal series of muscles, some of which emerge to form the lateral dilators of the esophagus. In the anterior part of the thorax the esophagus dips into the cardia as the esophageal valve. This valve is a deep curtain, thicker at base than at the free border. At the free border the space between the inner or direct face and the outer or reflected face is occupied by a blood sinus. Above, the space is filled by the circular fibers of the annular muscle (*ann m*). In young larvae the cells of the epithelium of the valve, except for a more columnar aspect at the curve at the shoulder of the cardia, upper bend, are like those of the epithelium of the hinder end of the esophagus. Immediately beyond the "upper bend," the esophageal type of epithelium is succeeded by the type characteristic of the cardia and stomach. For convenience, the point where the change occurs will be called the "break" in the epithelium. In mature larvae a band of narrow columnar cells (*ann*) much like the adjacent cells of the reflected face of the valve, but staining more deeply and with more vacuolated cytoplasm, is differentiated on the valve side of the "break." This functions in the pupal stage as a regenerative ring.

Delicate longitudinal fibers pass from the esophagus across to the shoulder of the cardia (*fi*), outside the valve, as is the case with several other Dipterous larvae, but the large blood sinus found in the larva of Simulium at this point (Miall and Hammond, : OO) is poorly represented in Culex and Anopheles. Miall and Hammond also describe the fibers in the larva of Simulium as muscular. My sections and dissections of Simulium do not lend themselves to this interpretation. The fibers are evidently not muscular in Culex and those of Simulium appear to me to be similar connective elements from the sheath which encloses both esophagus and cardia. If the similar fibers in any form prove to be muscular they must not be thought of as representing the longitudinalis muscle-coat. For in Dicranota (Miall, '93) which is exceptional in having the longitudinal muscles developed over the shoulder of the cardia, these pass beneath the blood sinus. The fibers are in the usual position external to the sinus. In Simulium, Miall has observed a distribution of the intima wholly like that observed in the fresh gut of Culex. The esophageal valve of Anopheles closely resembles the valve of Culex, but has a band of longitudinal muscles within the valve between the annular muscles and the epithelial cells of the upper part of the reflected face.

The salivary duct of the Culex larva is an extremely thin-walled tube which is not differentiated at any point. It is lined with a distinct chitinous intima. It forks as the occipital foramen is passed and either branch enters a cylindrical gland. These glands taper somewhat at their proximal or duct ends. The lumen is ample and not infrequently a distension occurs at the distal end of the gland which appears in the living larva as a spherical body of high refractive index. The epithelial cells of the living gland are almost transparent. In sections the cells are clearly defined, their cytoplasm stains deeply, the moderate sized nuclei (11.5 $\mu$) are reticulate, and nucleoli are present. There are usually not more than five or six cells in the cross section at any point. As the wriggler nears the pupa stage numerous vacuoles occur in the cytoplasm of many of the gland cells, so that it appears spongy. This is a sign of degeneration. Immediately prior to pupation, a belt of small cells forms at the neck where the gland narrows to meet the salivary duct. The cells of this belt are seemingly derived from the older cells of the neck of the gland and perhaps also from the epithelium of the outer end of the duct. They give rise to the imaginal salivary glands.

*The midgut.*— The midgut (pl. 16, figs. 37, 45) of the larva shows three regions: cardia, a ring of eight caeca, and the stomach. The cells of the cardia stain more deeply than those of the caeca or stomach epithelium, but are otherwise similar. As already noted, the fore gut intima terminates at the "break" where the cardiac epithelium begins near the upper bend of the esophageal valve. The cardia is lined by the peritrophic membrane. In the mosquito the cardia of the larva gives rise to the cardia of the imago through a metamorphosis similar to that by which the stomach of the larva gives rise to the stomach of the adult. The esophagus and esophageal valve on the other hand, have a wholly different metamorphosis, so that in Culex the cardia would appear to be unquestionably a part of the midgut. This seems to be the case in Nematocerous flies generally, although the evidence is as yet not complete for any one form. The larva of Anopheles, Chironomus, and Simulium like that of Culex has a cardia whose epithelium resembles that of the stomach. The embryological development of the cardia of the Chironomus larva shows it to be of mesenteric origin in this case (Miall and Hammond, : 00). Its relation to the cardia of the imago has not been studied. From Vaney's (: 02) account it would seem that the larva of Stratiomya belongs to this type and the cardia, "proventricule," of the larva of Ptychoptera as figured by van Gehuchten ('90) is more like stomach than esophagus in the histological structure of its epithelium. Occasionally no cardia is developed and in such instances the esophageal valve hangs directly into the stomach, e. g. Phalacrocera (Miall and Shelford, '97) and Dicranota (Miall, '93).

The cells of the epithelium of the eight caeca are transparent, polygonal, subequal in size, and show considerable intercellular substance. The nuclei are large and have nucleoli. Although the surface of these pouches appears nodulated as if from the contraction of muscle fibers, none were found. Sections show that the cells are on the average larger than the cells of the cardia or mid-intestine and very irregular in depth. The caeca usually contain a dark ochre-colored fluid.

The stomach is cylindrical and extends back to the fourth or fifth segment of the abdomen. It is abundantly supplied with tracheae and the cells of its walls are polygonal, with granular, yellowish cytoplasm, large oval nuclei ($17 \times 10\ \mu$), and nucleoli. The cells vary greatly in size, but are of uniform depth. Like the cells of the cardia and caeca they have a striated border. The walls of the stomach are nodulated, but this seems to be due more to the irregular size of the component cells than to the contraction of muscle fibers, although scattered muscles are visible in sections. In preparations, the basement membrane is very distinct. Protrusions of cells or parts of cells into the lumen, either in fresh or sectioned material was not observed for any part of the midgut.

The above description would not be wholly applicable to the stomach and caeca at all stages. For the processes by which the epithelium is destroyed and replaced during the pupa stage begin before larval life is ended, and involve progressively more and more cells as the wriggler nears the moment of pupation. Small regenerative nuclei appear in increasing numbers between the bases of the cells of the epithelium. The cytoplasm near the free margin of the cells becomes spongy. The striated border dissolves in places and occasionally cells protrude into the lumen of the gut.

*The hind gut.*— In the neighborhood of the fourth and fifth segments of the abdomen the stomach narrows and the hind gut begins. This portion of the alimentary canal is divided into ileum, colon, and rectum. The ileum (pl. 16, fig. 45, *il*), extends back to the eighth joint of the abdomen and is not separated from the stomach by a valve. Probably this is in adaptation to the continuous passage of bulky wastes through the gut. In the imago, in which the meal of blood is stored in the stomach while it is being digested, a valve exists between the ileum and stomach. The freshly extracted ileum appears as if wholly composed of circular muscle fibers, is transparent, and has an intima which is thrown into longitudinal folds. Sections demonstrate a very thin epithelium, the nuclei of the flattened cells measuring 12 by 6.5 $\mu$. Longitudinal muscles cannot be found. For a space immediately behind the stomach the circular muscles are also wanting (pl. 16, fig. 45, $x$), and into this naked annulus the five Malpighian tubules empty. These tubules in every way resemble the Malpighian tubules of the perfect insect.

The second region of the hind gut, the colon (pl. 14, figs. 26, 27, $co$) is distinct from the ileum, but is not sharply marked off from the rectum. Raschke ('87) grouped colon and rectum together as "Enddarm" and Hurst's figures include both as "rectum." In the text, however, the colon would seem to make part of his "ileum." The colon lies wholly within the eighth abdominal joint. Its walls have a deep epithelium of large polygonal cells and appear to be rather unyielding. Belts of slender circular muscles occur at regular intervals, but no longitudinal muscles can be demonstrated. The living cells of the colon are granular. Sections show faint striations radiating from the region of the nucleus. The oval nuclei of the cells measure 12 by 16 $\mu$.

Posteriorly, the colon passes gradually into the rectum (pl. 16, fig. 44, $rec$) which is wholly within the last joint of the abdomen. The rectum has delicate walls that tear easily. Sections show a very thin epithelium, resembling in structure the epithelium of the colon, and a series of regularly placed, slender, circular muscles.

### THE PUPA.

The first absolutely diagnostic sign of the on-coming of the pupation moult as observed in *Culex stimulans* is the appearance of two white spots in the prothorax, as the air chambers at the base of the pupal respiratory trumpets fill with air. At this moment probably the trumpets are evaginated beneath the cuticle. The larva becomes increasingly turgid in aspect and intermittent peristalsis-like waves of contraction pass over its body. The turgescence is largely due to the evagination of the pupal thoracic appendages under the cuticle. The charging of the air-sacs usually precedes the moult by about three minutes, but in one instance eight minutes elapsed, during almost all of which time the larva lay passive.

Suddenly a crack darts across the top of the head and the pale pupal head shows in the gap. From two to fifteen seconds later the cuticle of the thorax ruptures along the mid-dorsal line and the respiratory trumpets spring up to the surface of the water. The tear in the thoracic cuticle widens and the integument of the anterior abdominal rings also splits along the back as the pupa rapidly works the head and thorax clear and wriggles forward in the compressed remains of the old skin, until only the apices of the mouthparts and legs, and the tip of the abdomen are still sheathed by it. The abdomen is straight and each ring lies forward of its counterpart in the old cuticle. Hence the intima which is being withdrawn from the main longitudinal tracheae stretches back from the eighth segment to the respiratory siphon like a white cord on either side. Finally the tracheae are cleared, and the liberated pupa slips forward with flexed abdomen, rests for a few seconds, and darts away, perfectly formed, but of a light maltese color. The moult from the moment when the head cuticle splits to the clearing of the tracheal trunks has occupied from one minute five seconds, to one minute forty-five seconds. The whole process from the appearance of the air-sacs to the final darting away of the pupa takes on an average four minutes, but may occur in as short a time as two minutes or be prolonged beyond ten minutes.

The external aspect of the Culex pupa with its flattened head, high-arched thorax and nine-jointed abdomen is familiar. Male pupae can be distinguished from female by the larger gonapophyses. The gray color of the new pupa changes in an hour or so to the typical black brown. A few hours before the fly emerges the pupa becomes quite black, owing to the formation of the imaginal scale pattern under the integument. Shortly after pupation a space is formed between the integument and the dermis of the pupa (Hurst) into which the scales and hairs project. This fills with air before the fly emerges but the imago does not become much displaced in relation to the parts of the pupa skin. No violent movements precede or accompany the liberation of the fly. The abdomen of the pupa straightens and the imago quietly works itself free by slight contractions, the emergence on the average occupying four or five minutes. As much more time may elapse before the insect takes wing. I have found in Corethra and Chironomus that the space between the cuticle and the dermis fills with air prior to the emergence of the fly as it does in Culex. I have observed the liberation of the imago of Chironomus only. Here a large amount of air is admitted under the skin, so that the pupa has a glassy appearance, the preliminary movements are prolonged and violent, and the imago becomes markedly displaced in relation to the pupal parts. The apex of the abdomen, for example, finally lies three joints cephalad of its own sheath. In one species studied, the small apple-green fly emerges so rapidly that it takes wing within fifteen or twenty seconds from the moment when the pupal cuticle first ruptures. Comparing this with the slow emergence of Culex leads me to think that a connection exists between the rapidity of the emergence, the amount of air admitted beneath the cuticle, and the relative displacement of the imago in relation to the pupal sheaths.

*The mouthparts.* — The mouthparts of the pupa develop during the last larval instar. The two organs that arise from invaginate imaginal buds, the labium (pl. 15, figs. 32, 34, *li*) and the maxillary palpus (pl. 15, fig. 29, *maxp bd*), appear first. The development of the labial bud may be taken as typical as regards the histological changes. The dermis beneath the labial sclerite thickens and proliferates, until a plate of minute cells, which stain deeply with haematoxylin, is formed. While steadily growing upward, this plate is depressed and finally the bud consists of a bipartite eminence arising from the floor of a deep pocket. Both pouch and eminence secrete a delicate cuticle. The labial imaginal buds of Chironomus, Anopheles, and Corethra closely resemble the bud in Culex and a similar double structure has been described for the Pupiparous fly, Melophagus (Pratt, '93). Later in the instar than the labial and palp buds, the dermis of the top of the head thickens, stains deeply with haematoxylin and grows out as a cylindrical pouch. This bud forms the labrum (pl. 15, fig. 32, *la bd*). It projects caudad, depresses the adjacent tissues and lies in a shallow mesial furrow. Of slow growth until immediately before pupation, it then elongates rapidly, and reaches back to the origin of the epipharyngeal muscles (*epi m*) before the head cuticle splits. At the same time that this bud elongates, the cells of the hypodermis of the flabellae enlarge and show a marked affinity for "orange G." The dermis then retracts, separating from the overlying cuticle. In terms of their dermis the flabellae are converted into shallow

depressions or pits in the wall of the head. The dermis of the palatum remains in position. The mandibles and maxillae undergo
almost no metamorphosis prior to pupation. Shortly before the
moult, their dermis thickens and takes on some of the staining
reactions of the dermis of an imaginal bud, assuming a condition
intermediate between the cells of the flabellal dermis and the cells
of the labral, labial, or palp buds.

The larvae of Anopheles pass through a similar series of changes
as far as their mouthparts are involved, so that the wrigglers of
both Culex and Anopheles have at maturity an evaginate labral bud,
invaginate buds for the labium and the maxillary palpi, thickened
mandibular and maxillary dermis, and the dermis of the flabellae
and epipharynx retracted from the cuticle. The changes beyond
pupation are not known for Anopheles. With Culex, the successive alterations during the pupation moult are shown in figures 48
–50 on plate 16. As the head cuticle is sloughed, the labral bud
swings over into the plane of the other mouthparts. Probably
its rapid growth just prior to the moult is an important factor in
rupturing the head cuticle, while the change of position during
pupation must assist in pushing the head case away at an angle
favorable to the extraction of the ventral mouthparts (pl. 15, fig.
36). The change of position is partly due to the growth of the
labrum, partly to a contraction of the retractors of the flabellae, and
partly to a shortening of the dorsal wall of the buccal cavity. The
moult reduces the retractor muscles to ovoid masses (pl. 17, fig. 56,
retr fl), and shifts the flabellal depressions back to the entrance of
the pharynx. The mandibles and maxillae are withdrawn as cylindrical tubes. The bud of the maxillary palpus elongates slowly at
first and the labial bud has a still more tardy evagination. This is
probably to allow time for the retraction of the transverse fold
formed by the hypodermis of the transverse line, mental sclerite,
and the front wall of the pocket of the bud (pl. 15, figs. 34, 36,
fold).

The mouthparts of the pupa are closely pressed together (pl.
14, fig. 24) and form the central part of an ovoid shield which
covers the front and sides of the body. The antennae, legs, wings,
and halteres successively build the lateral parts of this shield. The
shield is in close contact with the body along its borders posteriorly
and ventrally, and above in front the central part is fused to the

thorax. Below in front and generally along the sides of the body an irregular space is left between body and shield. This space is filled with air (Hurst) and into it the recurved apices of the mouth-parts and the folded legs project.

When the dermis shrinks away from the cuticle a few hours after pupation, the various trophi become oval or cylindrical tubes and remodeling of the imaginal forms is possible. The mandibles, maxillae, and maxillary palpi — these last occupy the same cuticular sheath as the maxillae — undergo the least modification. In male pupae the mandibular sheath was empty by the end of the first day of pupal life. During a variable period of not less than ten hours, the labrum remains circular in section. Then its ventral face becomes infolded, and through differentiation of the furrow, the proboscis canal or so called epipharynx is formed (pl. 14, fig. 24, pc). The labium retains a cylindric or oval cross section until near the end of the first day of pupal life. Then the dermis of the mid-dorsal face is either elevated as a low crest or proliferates and thickens (pl. 14, fig. 24, hyp). In both sexes cells appear in the cavity of the labium beneath the altered wall and ultimately build the salivary gutter. These cells probably arise by invagination or migration from the modified area above. After they appear, the labium of the male pupa loses the mesial differentiation and quickly moulds itself to the imaginal form, but with the female pupa the ridge or thickening finally separates from the labium as the hypopharynx (pl. 14, fig. 25). I did not succeed in obtaining sections which showed the actual separation. After disassociation, both labium and hypopharynx quickly attain the imaginal structure.

*The fore gut.* — Since the mouthparts of the pupa are in close contact with one another, it cannot be said that the alimentary canal is lengthened by the addition of a proboscis canal so long as the dermis lies against the cuticle of the stylets, and although represented by the intracuticular spaces between labrum and labium after their dermis has parted from the cuticle, yet the canal is not actually formed until the ventral wall of the labrum is invaginated. If a new-formed pupa be examined, the fore gut will be found empty and clean. The buccal cavity is closed by the collapse of its walls, except at the posterior end, beneath the shallow dorsal diverticula which represent the flabellae and epipharynx (pl. 15, fig. 36, *epi*). Structureless *débris* from the strands of protoplasm that formed the cores of the epipharyngeal and flabellal setae fills the diverticula and spreads into the buccal space. In front of the curve where the potential proboscis canal enters the buccal cavity, the salivary duct opens. The pharynx retains something of the form which characterized it in the larva and the less markedly columnar epithelium as contrasted with the epithelium of the buccal cavity. The roof has evidently shortened, since the dorsal pharyngeal muscles are crowded toward the elevators of the palate, but this and other changes in the relation of organs cannot be accurately formulated, owing to the lack of fixed points from which to measure.

After the third hour of pupal life (pl. 17, fig. 56) the lumen of the buccal cavity opens again. It is soon freed from detritus and remains clean during the rest of the metamorphosis. The flabellal pits steadily diminish in size and finally fade out. The cells of the epipharyngeal pit histolyze and before the tenth hour is passed have dissolved into a mass of broken elements. This lies within the cavity of the head, completely cut off from the buccal cavity (pl. 17, fig. 57, *epi*), and is slowly absorbed without the intervention of phagocytes. Seemingly the separation of the degenerating epithelial pouch from the buccal cavity (pl. 16, fig. 39) is accomplished by an inpushing of the adjacent epithelium rather than by a proliferation of cells at the margin of the pit. Mitotic figures are absent. It is not impossible, however, that some of the epithelium which closes beneath the degenerating cells and cuts them off from the lumen of the buccal cavity is derived mediately or immediately from the epithelium of the epipharynx as it is destroyed. If this is so, and the epipharynx of the mosquito wriggler is a true epipharynx, part of the anterior hard palate of the imago is epipharyngeal in origin. And in such a case the small conical setae near its apex might gain a morphological value as epipharyngeal structures   This will not affect the character of the labrum as a simple and not a compound organ.

While the changes described have been taking place, the pharynx has become cylindrical and the cells of its epithelium now exactly resemble the cells of the epithelium of the buccal cavity. It is not clear how this alteration is accomplished. No mitotic figures can be found and no waste is given off into the lumen of the gut. The loss of histological and structural distinctions between the buccal cavity and pharynx leaves as landmarks for the entrance to the pharynx (pl. 17, fig. 56, *) only the cingulum muscle below and the anteriormost dorsal pharyngeal muscle above.  The dorsal mark is open to the objection that this muscle lies behind the actual boundary; and the relations in this neighborhood are obscured also by the retraction of the flabellal hypodermis back to the epipharynx — which does not seem to have moved much — and by the shortening of the roof of the pharynx.  These criteria enable us to judge the morphological value of any comparison between the limits of the larval and imaginal fore gut regions.  Nevertheless, it will be interesting and perhaps not unprofitable to work these out (pl. 17, fig. 56-58).  Roughly speaking, that part of the pupal buccopharynx which is in front of the anteriormost dorsal pharyngeal muscle and the cingulum muscle represents the larval buccal cavity. Ventrally, the boundary early becomes indistinguishable owing to the loss of the cingulum muscle, but dorsally a slight bend comes in near the point where the two regions meet (*).  This bend coincides with the rear border of the soft palate of the imago. Pharyngeo-esophageal boundaries are not developed in the larva ventrally, but dorsally the insertion of the retractors of the pharynx (retr ph) marks the rear of the pharynx.  These muscles metamorphose into the ascending pharyngeal muscles of the imago (asc ph). The anterior dilator muscles immediately caudad become the anterior dorsal dilator muscles of the antlia of the imago.  The epipharyngeal pit lies at first above and then forward of the marked curve in the gut.  This curve remains after the epipharynx and its muscles have vanished and the front wall of the clypeus forms distad of it.  So, unless it shifts position markedly, for which there is no evidence, the epipharyngeal muscles of the imago insert on the roof of the buccal cavity very near the point occupied by the larval epipharynx.

The esophagus undergoes no metamorphosis beyond the destruction and regeneration of the muscle-coats, the differentiation of the antlia at its anterior end (pl. 17, fig. 57), and certain changes which involve the esophageal valve.  The esophageal epithelium is handed intact to the imago, probably with an increase in the number of the component cells.  The muscle-coats histolyze about the eighth hour and are replaced somewhat later by the imaginal muscles.  The process could not be studied in detail because of the minute size of the new elements.   The antlia gradually forms during the first half of the second day of pupal life, but in common with other parts of the alimentary canal the characteristic imaginal intima with its division into plates is not developed until just before the fly emerges. This intima is secreted beneath the delicate undifferentiated pupal intima which lines the stomodaeum.   The proctodaeum similarly has a pupal intima beneath which the imaginal intima is secreted. The lumen of the esophagus remains clean during the metamorphosis, except at the posterior end, where granular rubbish containing chromatin is found, seemingly drifted in from the cardia.

All the muscles of the head continue intact until the eighth or tenth hour and then histolysis occurs.   But although small leucocyte-like cells can occasionally be noted, these are rare and there is nothing which definitely assigns a phagocytic *rôle* to them. The mandibular muscles, the depressors of the antennae, the muscles of the maxilla, the epipharyngeals, the retractors of the flabellae, the diagonals, the dorsal pharyngeal muscles, the ventral retractors of the pharynx, and the cingulum histolyze and are absorbed between the tenth and thirtieth hour of pupal life.   The first imaginal muscles appear about the eighth hour with two bands of small dark myoblasts, probably the maxillo-labial muscles.   By the seventeenth hour the sides and floor of the head cavity are traversed by similar belts of small myoblasts.   The ultimate source of the new cells is uncertain, owing to the minute size of the elements of both the larval and the imaginal muscles.   From the moment of their appearance the new nuclei are closely associated with the dermis of the head and their cells resemble the small, dark cells of this epithelium.   There is, however, nothing which would militate against applying van Rees' ('88) interpretation of the muscle metamorphosis of Musca to Culex.   By this the new myoblasts would be derived from the nuclei of the larval muscles.

The retractors of the pharynx, the elevators of the dorsal plate, the lateral muscles, and the anterior, posterior, and lateral dilators of the esophagus are worked over into imaginal muscles.   The anterior esophageal dilators alter first and their metamorphosis may be taken as typical.   A majority of the large reticulate nuclei become smaller, darken, and lose the reticulum ; others histolyze. Then a large number of small dark nuclei appear, perhaps derived from the older muscle nuclei by amitotic division.   The two sets of nuclei, older and younger, remain side by side until the twentieth hour is reached, when the contractile substance, hitherto hostile to haematoxylin, stains readily with it and all the nuclei are larger and reticulate — the imaginal condition. The final increase in the amount of the contractile substance and the complete attainment of the imaginal staining reaction are soon reached. The relation of the larval to the imaginal muscles is as follows: —

| Larval Series. | Imaginal Series. |
| --- | --- |
| Retractors of the pharynx. | Ascending pharyngeals. |
| Lateral muscles of pharynx. | Lateral pharyngeals. |
| Elevators of the dorsal plate. | Muscles of the valve. |
| Anterior dilators of the esophagus. | Anterior dorsal dilators of antlia. |
| Posterior dilators of the esophagus. | Posterior dorsal dilators of antlia. |
| Lateral dilators of the esophagus. | Lateral dilators of antlia. |

Probably only the posterior portion of the muscle complex that I have called the lateral dilator of the esophagus, gives rise to the lateral dilator muscles of the imago's antlia, the anterior portion being histolyzed. The hypopharyngeal muscles of the adult are derived from the fibers beneath the hypopharyngeal sclerite. Those muscles dorsad of the alimentary canal which have forerunners among the larval muscles appear from the twenty-fifth hour onward and the clypeus begins to differentiate at this time (pl. 17, fig. 58).

The esophageal valve of the new-formed pupa is partly drawn out of the cardia, so that the annulus of regenerative cells (*ann*) lies in line with the walls of the cardia and the upper bend of the valve has been shifted to a point low down on the reflected face. The regenerative annulus also has widened, probably through modification of adjacent cells of the epithelium of the valve. In a pupa five hours old (pl. 16, fig. 38) the valve is completely withdrawn from the cardia. The regenerative annulus is very prominent and the annular muscle of the valve (*ann m*) has begun to degenerate in advance of the histolysis of the muscle-coats generally. An hour later, the regenerative annulus is represented by a low circular ridge of columnar cells which projects into the lumen of the esophagus just above the cardia. The valve lies immediately cephalad as a low ring-fold of the wall of the gut, girdled externally by the remains of the annular muscles. By the tenth hour (pl. 16, fig. 43) the parts are hardly recognizable; for the valve (*v*) is flattened and almost obliterated while the cells of the annulus (*ann*) have largely lost their characteristic histological structure and resemble cells of the adjacent esophageal epithelium. The eleventh-hour pupa has almost the imaginal relations (pl. 16, fig. 40). The regenerative ring and larval valve have disappeared, the epithelium is everywhere of esophageal type and near the cardia the walls are folding inward to build the imaginal esophageal valve. The rapidly diminishing belt of *débris* which remains from the degenerated annular muscles lies cephalad of the new valve. The infolded region clearly is epithelium that at one time formed part of the regenerative annulus.

At about the tenth or eleventh hour the esophageal diverticula push out just caudad of the histolyzed annular muscles. They probably do not involve cells that have formed part of the regenerative annulus, but arise from esophageal epithelium which has not been altered (pl. 16, figs. 40, 43, *f res*). The out-pushings quickly enlarge to small pouches, the ventral one opening widely into the lumen of the gut, the dorso-lateral ones with narrow stalks. Growth is then slow until the thirtieth hour of pupal existence, after which the pouches rapidly attain the respective imaginal proportions. The epithelium remains like that of the esophagus, columnar, till the very end of pupal life, when the characteristic imaginal epithelium is developed, seemingly by a flattening of the cells, accompanied by a folding and wrinkling of the walls. I am not certain whether these pouches secrete a pupal as well as an imaginal intima. When the mosquito leaves the pupa case the sacs are empty, but air bubbles appear shortly after emergence.

It is not possible to escape the feeling that a relation exists between the formation in Culex of the annulus of regenerative cells close to the cardia and the development of the "anterior imaginal ring" at the same point during the metamorphosis of other flies. Such a ring of regenerative cells has been described for Chironomus, Anthyomyia, Stratiomya, Tanypus, Gastrophilus, and Musca (Vaney, :02). Its fate has been determined for Gastrophilus and Musca. In the former the whole stomodaeum is destroyed and is replaced "entirely by proliferation of cells derived from the buccal discs and the interior imaginal ring" (Vaney, :02). With Musca the stomodaeal epithelium is partially destroyed at least and is rebuilt from scattered regenerative cells, as well as from the imaginal ring (van Rees, '88). In Culex the esophageal epithelium seems to suffer no loss of cells and the regenerative annulus gives rise to the imaginal esophageal valve by simple remodeling.

The salivary glands frequently begin to degenerate before pupation occurs and the destructive processes advance rapidly during the first hours of pupal life. The cytoplasm of the epithelial cells becomes more and more spongy, large vacuoles appear here and there, and the nuclei decrease in size. Meanwhile the small regenerative cells at the neck of the gland are steadily increasing in number. By the twentieth hour the glands have shifted to a more ventral position and the regenerative epithelium stains deeply with haematoxylin, a characteristic of regenerative epithelium generally. The nuclei of the gland cells are now still more diminished in size and the vacuolization is extensive, especially near the belt of regenerative cells. Frequently cells are so distended by vacuoles that they obliterate the lumen of the gland. About this time the imaginal salivary glands appear as three small cylinders developed from the regenerative epithelium. As soon as these form, the larval glands histolyze completely, become a mass of granular detritus filled with fragments of chromatin and are absorbed without phagocytosis in five or six hours. In their places on either side of the thorax lie the three slender tubes of the imaginal glands. Just before the mosquito leaves the pupa case the glands enlarge, shift to the adult position beneath the alimentary canal, and the central and lateral acini become unlike. Whether the final increase in size is due to increase in the number of component cells or only to growth of cells already present, could not be determined. After the emergence of the fly, the nuclei of the cells enlarge slightly, and the cells become distended with the stored salivary secretions until three times their former dimensions.

The epithelium of the cardia and stomach is totally histolyzed and a new epithelium is formed from regenerative nuclei which appear among the older cells. This course of development has been observed for the midgut of all flies whose metamorphosis has been studied (Korscheldt and Heider, '99). The larva of Culex reaches pupation with degenerative and regenerative processes well established. The cytoplasm of the epithelial cells is vacuolated, producing a spongy appearance, the striated border is dissolving, and occasionally cells protrude into the lumen, of the canal. Between the bases of the older cells the regenerative nuclei occur in increasing numbers. With the larvae of *Tenebrio molitor* (van Rees, '88) the regenerative nuclei are found even in the first instar. I do not know how early they appear with the mosquito wriggler. In the last larval instar and the newly formed pupa they are very numerous. As these nuclei increase in number, they diminish in size, which suggests that they are derived by repeated divisions from those present in the larval stage.

The changes advance swiftly after pupation. In a pupa one hour old (pl. 14, fig. 28) the epithelial cells are separating from one another, are elongate and protrude into the lumen, the striated border has been everywhere lost, and the cytoplasm is dissolving. By the third hour of pupal life, the regenerative nuclei, which are now about the size of the epithelial cells of the gut of the imago — less than one half as large as the regenerative nuclei of the larva — have forced the older cells from the basement membrane and form a definite pavement epithelium with recognizable cytoplasm. The older cells lie in fused masses in the lumen of the gut, their nuclei reduced to dark granulate spheres. Here they are rapidly absorbed (pl. 16, figs. 40, 43, *ca*). This metamorphosis does not proceed with equal celerity in all regions of the midgut. The caeca, the cardia, and the iliac end of the stomach lag behind the major part of the stomach. The caeca decrease in size as their epithelium histolyzes, and by the fifteenth hour they have vanished. The epithelium of the cardia degenerates first at the posterior end of the chamber. The new epithelium, unlike that of the caeca and the stomach, consists of columnar cells from the beginning and appears to be formed as an advance forward beneath the older cells (pl. 16, fig. 38, *ca*), instead of simultaneously over all parts of the wall.

At the beginning of pupal life the hind gut regions, ileum, colon, and rectum are the same as in the mature larva, except that they are shortened and that a ring of small dark cells lies between the rectum and the anus (pl. 17, fig. 54, *r an*). This ring develops during pupation and the first hour of pupal life, but whether its cells are derived from the cells of the posterior end of the rectum or from cells of the hypodermis in the region of the anus could not be determined. The rectum and ileum are involved in degenerative

and regenerative processes which run the same course in all parts
of both regions, but could only be studied in detail for the ileum.
The metamorphosis of the colon is very peculiar.

According to Hurst ('96) the epithelium of the intestine "divides
into a thin outer and a thick inner layer. The latter becomes loos-
ened, breaks up and appears to be digested." The process for the
species of Culex that I studied is not as simple as this would imply.
The flattened epithelium first becomes columnar. Then the cells
degenerate, separating more or less from one another. Large vacuoles make their appearance, the cytoplasm becomes spongy, and a
small amount of *débris*, in which granules of chromatic material are
noticeable, is given off into the lumen of the canal. In all regions
considerable mitosis was observed toward the end of the changes.
The processes begin at the upper end of the ileum, run their course
here in about six hours and then involve simultaneously the rest of
the ileum and the rectum, lasting eight or ten hours. They result
in the formation of an epithelium of small, dark-staining cells, somewhat like the cells of the new anal ring. These gradually alter to
the adult characters for each region. Enlargement and probably
proliferation of cells in certain parts of the rectal walls form the
rectal papillae (pl. 17, fig. 53, *p*). The muscle-coats are lost about
the seventeenth hour and are later replaced by new muscles.

It is evident that in the metamorphosis of the ileum and rectum
certain of the epithelial cells are eliminated and new cells, derived
from epithelium which escapes destruction, replace them. Probably
the degenerative process involves all the older cells to a greater or
less degree. The nuclei of the reconstructed epithelium are barely
one half the size of the nuclei of the older cells. The nature of the
preliminary alteration by which the epithelium becomes columnar
is obscure. It does not seem, however, to depend on any contraction of the walls of the gut.

The alterations recorded above are not without relation to the
postembryonic development of the hind gut in such flies as Musca
or Gastrophilus. In the latter genus, the hind gut is destroyed and
is reconstructed from an anterior imaginal ring of regenerative cells
near the Malpighian tubules and a posterior ring near the anus
(Vaney, : 02). The hind gut of Musca evidently undergoes a similar metamorphosis, but observers do not agree as to the amount of
epithelium destroyed. Kowalevsky ('87) asserts that all the older

cells are swept away while van Rees ('88) believes that a considerable part of the older wall remains.   In Culex the appearance of the
regenerative changes in the neighborhood of the Malpighian tubules
before the remainder of the gut is involved, and the formation of
an anal ring of new cells — whatever their ultimate source — furnish interesting parallels to the conditions found in Musca and Gastrophilus.   The resemblances of course mean no more than the
presentation in varied degree of certain tendencies which exist
among the Diptera.   A closer genetic connection is not implied.

As an illustration of the improbability of such a connection we
may take the changes which occur in the colon of Culex.   The
development of this region recalls the reconstruction of the hind
gut in the more specialized flies from anlagen at either end of the
canal.   But the processes are not really comparable with those
found in Musca or Gastrophilus and when we consider the simple
metamorphosis of the remaining hind gut there is marked and
peculiar specialization.   The colon of Culex is destroyed and
replaced by a tube formed from a backward growth of the cells
of the ileum and a forward growth of the cells of the rectum.
The proliferation begins as soon as the iliac and rectal epithelia
are reconstructed, i. e., about the seventeenth hour.   At this time
the cells of the shrunken colon are spongy, and have separated
from one another.   Their nuclei have diminished in size but are
intact (pl. 17, fig. 52, co).   The invading cells enter the lumen (pl.
17, fig. 59), and active mitosis is observable at and near the head of
either advance.   The two cell armies seem to progress with equal
speed.   As soon as the advance is well under way, the nuclei of the
colon cells lose their reticular structure and reduce to masses of
chromatic material within the nuclear membranes.   Nuclei and
cells, however, remain distinct until individually displaced by the
invading cells, and then quickly histolyze.   In a nineteen-hour
pupa, cells already displaced lay outside of the new epithelium as
a mass of fragments, while cells still in position remained distinct
units.   By the twentieth hour of pupal life (pl. 17, fig. 53) the
new canal (ilc) is completed and the old colon (co) is represented
by débris.   This is quickly absorbed without the intervention of
phagocytes.   The muscles of the larval colon histolyze in situ.
The muscles of the new ileo-colon are developed from myoblasts,
probably from the ileum and rectum.

The Malpighian tubules pass from the larva to the imago without visible change. This absence of metamorphosis appears to be the rule for these tubules in the Nematocerous Diptera, if observations on Simulium, Chironomus, and Psychoda (Vaney, : 02) added to those for Culex, give a sufficient basis for generalization.

*Notes on the metamorphosis.* — The organs of Culex outside of the alimentary canal undergo a metamorphosis as simple in grade as that of the mouthparts or alimentary canal. A few points with respect to these organs may be noted. The dermis of the head and the body is passed directly over to the imago. The muscles of the thorax and abdomen also are seemingly not involved in any alterations. The wing and leg muscles are new with the pupa, the rudiments of the former being present beneath the hypodermis of the dorsum of the thorax during the last larval instar. My observations on the fat body accord with those of Vaney. "In the lower Diptera (Culex, Simulia, Chironomus) the fat elements are maintained in their integrity from the larva to the adult." In the fat body dorsad of the nerve cord in the abdomen, there can be noted during the pupal stage scattered cells which resemble leucocytes more than other cells of the body. These do not, however, appear to function in any manner.

The eyes of the larva form part of the compound eyes of the perfect insect. Hurst ('90) has given a good account of the evolution of the imaginal eye, which "consists in the addition of new elements at the edge [of the larval eye] which arise by direct modification of the .... epidermis around the margin of the eye, epidermis whose last function was to secrete the pupal cuticle." He notes that "corneal facets are never formed in the pupal cuticle" and that the ocellus of the larva is inconspicuous in the adult. I find that this ocellus degenerates during the pupa stage. The cells become vacuolated, the pigment agglomerates, and the whole structure sinks beneath the dermis. In the perfect insect each ocellus is hardly more than a small mass of black pigment at the rear of the compound eye.

Both Culex and Anopheles have a pair of rudimentary ocelli which do not seem to have been described. They are situated on the vertex of the head, caudad and mesad of the bases of the antennae. Each consists of a plate of enlarged dermal cells, lying immediately over a small ganglionic mass of spindle-shaped cells. A fine nerve connects this ganglion with the supra-esophageal ganglion.    The overlying cuticle is not modified (pl. 12, fig. 2 ; pl. 13, fig. 8, oc). These ocelli are not present in the larva, but develop during the pupation moult.   Probably they represent in a vestigial form the lateral members of the three ocelli found on the head of many Nematocerous flies.   If this is the case, no trace of the median member can be found.

During the pupal stage the nervous system undergoes gradual changes which are growth rather than a metamorphosis, increasing greatly in size and probably also in complexity.   The brain of the imago is relatively enormous and there are two huge optic ganglia united above to the supra-esophageal ganglion (pl. 13, fig. 9–11, o gang).   A small buccal ganglion lies above the pharynx, between the last pair of elevators of the palate and the valvular muscles. The larva, on the other hand, has a small, transverse supra-esophageal ganglion and an insignificant infra-esophageal ganglion (pl. 15, figs. 29, 30).   The swollen lateral apices of the supra-esophageal mass correspond to the optic ganglia of the imago (o gang).   The buccal ganglion is very small (pl. 15, fig. 32, b gang).   The thoracic ganglia of the imago are united to form a single elongate mass.   In the abdomen six ganglia are distinct and occupy the joints from the second to the seventh inclusive.   In the larvae, however, (pl. 14, fig. 26) there are three separate thoracic ganglia and the first eight abdominal joints are furnished with a ganglion apiece.   During the later larval instars, the ganglia of the first and eighth joints of the abdomen shift to the anterior borders of their respective segments, and before pupation occurs, the ganglion of the first joint enters the thorax (pl. 17, fig. 54).   Later, it fuses with the thoracic ganglia to form the thoracic ganglion mass.   Some hours after pupation, the ganglion of the eighth abdominal segment passes into the seventh joint (pl. 17, fig. 55) and ultimately fuses with the ganglion already present in the segment to form the sixth abdominal ganglion of the imago.

The imaginal buds share in the simplicity of the metamorphosis. With the exception of those for the antennae, none are deeply invaginated, and as far as the mandibles, maxillae, and the tail fans of the pupa are concerned, the formative dermis does not alter beyond the thickening which is the invariable preliminary to the formation of an imaginal bud.   The labium, maxillary palpi, thoracic limbs, respiratory trumpets, wings, halteres, and gonapophyses are developed from shallow invaginate buds. The labrum is formed from an evaginate bud. The thoracic and antennal buds are evaginated immediately prior to the pupation moult (pl. 16, fig. 48), the labial and maxillary palp buds after the larval cuticle is partly removed. The buds for the gonapophyses are situated on the ventral face of the ninth joint of the abdomen. The patches of altered hypodermis which represent in a rudimentary form buds for the tail fans of the pupa are formed low down on the sides of the eighth segment of the abdomen. In the species of mosquito studied by Hurst these buds are described as "plate-like bodies lying immediately beneath the cuticle of the larval siphon." They lie high up on the sides of the segment in the larva of Anopheles, and in Corethra are almost dorsal in position. These differences are probably correlated with the varied size and shape of the eighth abdominal segment in the three genera: wide and unincumbered in Corethra, narrower in Anopheles with a small dorsal respiratory siphon, and very narrow in Culex with a huge dorsal siphon. The tail fan buds are the last to appear in the Culex larva. The order for the development of the imaginal buds is: antennal and thoracic; labial and maxillary palp; labral; gonapophysial; mandibular and maxillary; tail fan.

The antennal buds are of especial interest. They are deeply invaginated into the head cavity, and when completed the base of the new antenna at the bottom of the invagination pocket lies behind the eye spot (pl. 15, fig. 29, *ant bd*). Miall and Hammond ('92) compare the type of antennal and eye development presented by Culex, where the eyes are maintained independently of the antennal invaginations, with that found in Chironomus or Simulium where invaginations give rise to both antenna and eye, and finally with the extensive invaginations of Musca from which eye, antenna, and the whole head are formed: the so called "brain appendages." On a basis of this comparison he places Culex at the bottom of a series, Simulium and Chironomus higher, and Musca at the summit. This arrangement is perhaps justified as a statement of progressive elaboration of a tendency to form eye and head from an invagination that is connected with the antennal bud. But it does not sufficiently accentuate the mechanical aspects of the phenomenon. Kellogg (:02) appears to me to express the case better when he says: "Whether an organ, as wing, leg, antenna, or mouthpart, shall begin as an invagination or an evagination of the derm is chiefly a matter of mechanical necessity or ease and of the radicalness of the metamorphosis." The "brain appendages" of Musca, the deep cephalic pouches of Chironomus, the · shallow cephalic pouches of Simulium, and the independent eye and antennal formation of Culex do not constitute a closely locked series.    The character and position of the invagination differ in each instance. The genera can be compared only with respect to the radicalness of the metamorphosis in each case, by which varied modifications of tendencies that exist throughout the Diptera are presented.    That mechanical necessities are more potent than relationship in determining the character of a bud is well shown by a comparison of the antennal buds of Culex, Anopheles, and Corethra.    The larva of the first genus has a broad, capacious head, but the head of the larva of Anopheles is laterally compressed, and that of the Corethra larva is markedly flattened from side to side.    The antennal buds in Culex are thrust back into the cavity of the head.    In the other genera the antennal buds are thrust back in deep longitudinal furrows on either side of the head, and the invagination pocket is narrowly open for its entire length.

## Summary.

The alimentary canal of the mosquito has a midgut and hind gut which are not subdivided into numerous or well defined regions, but the fore gut exhibits the maximum amount of differentiation that has been found among the flies.    It may be divided into proboscis canal, pharynx, antlia, and esophagus.    The pharynx and antlia are pumping organs.    Their suctorial action depends on the modification of the chitinous intima to form definite areas or "plates," which spring inward and occlude the lumen, to be withdrawn again by muscles. This duplication of the pumping apparatus places the mosquitoes in the diantlial class of flies, a group which probably includes all the Nematocera.    The Muscidae and others of the Cyclorrhapha are monantlial, having only the anterior or pharyngeal pump.    There is no secondary union of the pharynx with the walls of the head, so that it is not converted into a fulcrum.    In this region, as is the case with all other flies, the walls and floor are rigid and the dorsal roof constitutes the plunger of the pump. The pyriform antlia on the other hand has walls that are modified to form three racquet-shaped plates, one dorsal, two ventro-lateral. In the narrower anterior end of the pump, forward of the circumesophageal nerve-ring, the walls become unyielding, so that the dorsal plate alone retains a limited amount of motion. In correlation with the inactivity of this part of the pump the anterior dorsal dilator muscles are vestigial. The Anopheles mosquitoes, which otherwise closely resemble the members of the genus Culex in the structure of their fore gut, have a more cylindrical antlia, the anterior end of which is much less rigid, and the dilator muscles are well developed. A valve separates the pharynx from the antlia and a sphincter muscle surrounds the rear of the latter pump as it gives place to the esophagus. The esophagus is a thin-walled tube with flattened epithelium and poorly developed muscle-coats. The esophageal valve is small and immediately in front of it the three esophageal diverticula arise. The tracheal supply to the fore gut is scanty, but the number of branches which run to the muscles of the pharynx and antlia indicate great activity for both pumps.

Each of the two salivary glands is composed of three acini, the middle member being unlike the lateral in the character of its secretions. The ducts from these glands unite in the rear of the head and the resulting median duct empties into a cup-shaped chitinous pump at the base of the hypopharynx. This pump forces the saliva along the salivary gutter of the hypopharynx into the wound.

The two regions into which the midgut may be divided, the cardia and the stomach, are not sharply distinguished from each other. Their epithelium is low and columnar, the muscle-coats are represented by slender longitudinal and circular muscles, and caecal pouches are not developed. This last is a point of difference between Culex and Anopheles, as the latter genus seems to possess rudiments of such diverticula. The tracheal supply to both cardia and stomach is extensive and the abdominal portion of the stomach is capable of great dilatation, serving as a storage place for the meal of blood while it is being digested.

A prominent valve is developed at the posterior end of the stomach, opposed to a similar valve at the upper end of the first region of the hind gut, the ileo-colon. Between the two sets of valves the orifices of the five Malpighian tubules are situated. The two regions of the hind gut, the ileo-colon and the rectum, are distinguished mainly by the greater diameter of the latter and the six rectal papillae on its walls. The epithelium is uniformly flattened, the muscles are much like those of the stomach or cardia, and the tracheal supply is extensive. A thin chitinous intima lines the hind gut, and is differentiated to form a hirsute belt immediately behind the valve at the upper end of the ileo-colon.

The regions to the alimentary canal of the imago and larva are noticeably unlike in several points, and the high degree of adaptability of the Dipterous larva generally, has produced within the group marked variations in the morphology of the alimentary tracts in the young of closely allied genera. Hence in a sense, the morphology of the larval and imaginal gut are separate problems. In the larva of Culex or Anopheles the following regions can be distinguished: buccal cavity, pharynx, esophagus, cardia, stomach, ileum, colon, and rectum. Of these, the pharynx is not well marked off from the esophagus below and at the sides, the cardia is more distinct from the stomach than in the imago, but clearly a differentiation of its anterior end, and the colon and rectum pass gradually into one another.

The buccal cavity has columnar epithelium and an epipharynx is differentiated on its dorsal wall. The epithelium of the pharynx is flattened. This region has the heaviest intima of any part of the larval gut and is an organ of peculiar shape with an elaborate musculature. The buccal cavity of the larva corresponds to the posterior end of the proboscis canal and the anterior two thirds of the pharynx of the imago. The pharynx of the larva forms the posterior third of the pharynx of the fly, while the anterior end of the esophagus becomes the antlia. The esophagus has flattened epithelium, strong circular, and well developed longitudinal muscles. Esophageal diverticula are lacking, but the esophageal valve is more highly specialized than the corresponding valve of the imago. It is less modified, however, than the esophageal valve of the larva of Anopheles, the alimentary canal of which otherwise closely resembles that of Culex. The salivary glands of the mosquito wriggler are cylindrical and the salivary duct is undifferentiated throughout its length. It opens freely on a small area of heavy chitin in the floor of the buccal cavity, the hypopharyngeal sclerite. This structure is not homologous with the hypopharynx of the imago.

The epithelium of the cardia and stomach is very unlike that of the same regions with the perfect insect, being composed of large polygonal cells which have a striated border and large nuclei. Eight cylindrical caeca protrude in a circle just behind the cardia and there is no valve at the posterior end of the stomach nor at the cephalic end of the ileum. The muscle-coats are weak.

Of the three regions of the hind gut, the ileum has an extremely thin epithelium and well developed circularis muscles, the colon an epithelium of huge polygonal cells and slender belts of circular muscles at intervals, while the rectum has a very thin epithelium indistinguishable from that of the colon and similar belts of circular muscles. The hind gut is lined by a strong chitinous intima.

The metamorphosis of the mosquitoes as shown by Culex is of low grade compared to that of other flies. No phagocytosis has been observed. The fat body, the Malpighian tubules, the ventral thoracic muscles, and the muscles of the abdomen seemingly are passed on intact to the imago. The eyes of the larva are the eyes of the pupa and are elaborated to form part of the compound eyes of the adult, the remainder of these organs being developed from the surrounding hypodermis. The larval ocelli degenerate. The nervous system increases greatly in bulk and complexity, two of the abdominal series of ganglia change their position, a Johnston's organ is formed within either antenna, and two vestigial ocelli appear on the vertex of the head. These changes are rather of the nature of growth than metamorphosis. The dermis of the head and body moulds itself to the pupal and imaginal contours and secretes the successive pupal and imaginal cuticular structures without visible mitosis or histolysis. The epithelium of the fore gut alters greatly in appearance, the pharyngeal and antlial pumps are differentiated, the larval esophageal valve fades away and a new valve is formed by the out-folding of the gut walls, and the esophageal diverticula are pushed out. Apparently the epipharynx alone undergoes histolysis during these changes.

No true imaginal buds are developed for the mandibles, the maxillae, or the tail fans of the pupa. The hypodermis merely thickens and stains more deeply. The labrum develops as an evaginate pouch of altered hypodermis on top of the head. The labium, maxillary palpi, thoracic limbs, wings, halteres, and gonapophyses are developed as invaginate buds. But the invaginations are very shallow in most cases and in the deepest of them — those for the antennae — the connection with the surface of the head is a wide channel. During the pupal stage the hypodermis of the mid-dorsal face of the labium thickens and forms a low ridge, beneath which cells build the salivary gutter. In male pupae the ridge disappears as soon as the salivary gutter is formed. In female pupae it separates from the labium and becomes the hypopharynx.

The longitudinal and circular muscles of the walls of the alimentary canal and most of the head muscles are completely histolyzed. It seems likely, however, that the head muscles and the muscles of the walls of the gut in the perfect insect are derived from myoblasts which are descendants of nuclei of the older muscles. A few of the muscles in the head undergo a partial degeneration only and in a sense give rise directly to imaginal bands. The muscles of the wings and legs of the imago are new with the pupa stage. The rudiments of the former series appear during the last part of the larval period.

The epithelium of the midgut, i. e., of cardia, caeca, and stomach, is sloughed and is replaced by a new epithelium derived from scattered regenerative cells. These appear during the larval stage, multiply rapidly, and during the early part of the pupal stage form an epithelium which displaces the older cells. The latter fall into the lumen and are absorbed. The larval colon is completely histolyzed and a canal which grows into its lumen by proliferation of cells of both ileum and rectum replaces it. Prior to this growth, the epithelium of the ileum undergoes a partial degeneration. Hence the posterior end of the imaginal ileo-colon is a new structure. The anterior end consists mainly of remodeled cells of the older epithelium. In the rectum the anterior portion has a metamorphosis similar to that of the ileum. The posterior end is rebuilt from an anal ring of regenerative cells the origin of which is uncertain.

The metamorphosis of the alimentary canal of Culex presents parallels to the postembryonic development of the most specialized flies. Thus, a belt of differentiated cells forms near the cardiac end of the fore gut, recalling in its position the anterior imaginal ring of Musca, from which much of the foregut of the imago is derived. In Culex the ring is remodeled to form the esophageal valve. The anal ring already noted, from which part of the rectum comes, and the fact that the degenerative and regenerative processes in the epithelium of the ileum commence and are completed near the Malpighian tubules before they involve the rest of the hind gut, recall the positions of the anterior and posterior imaginal rings in the postgut of many flies. Even the peculiar metamorphosis of the colon may be compared with the formation of the hind gut in the specialized Cyclorrhapha by a forward growth from the anal and a backward growth from the Malpighian regenerative rings, whether the older epithelium is destroyed completely as in Gastrophilus, or partly as in Musca. The alterations of the midgut also are precisely the same as those which are found in all flies, even the most specialized.

Moreover, comparisons may be made for other organs than those of the alimentary canal. For example, on the basis of the antennal buds, a series of genera may be formed, beginning with Culex where the larval — and pupal — eye is independent of the antennal bud, through flies where the eyes develop from the same invaginations as the antennal buds, up to Musca, a species in which the whole head is formed from a pair of large invaginations. But all such comparisons must be taken to mean no more than this, that Culex and the other flies present in varying degree and under varying conditions, tendencies common to the insect group of which they are representatives.

## LITERATURE.

Annett, H. E., Dutton, J. E., and Elliot, J. H.
: **01.** Report of the malaria expedition to Nigeria. Part 2. Liverpool sch. tropical med., mem. 4, p. 73-89, pl. 16-19.

Becher, E.
'**82.** Zur kenntniss der mundtheile der Dipteren. Denkschrift. acad. wissensch. Wien, vol. 45, p. 123-162, pl. 1-4.

Christophers, S. R.
: **01.** The anatomy and histology of the adult female mosquito. Reports to the malaria comm., royal soc. London, sec. 4, 20 pp., 6 pls.

Dimmock, G.
'**81.** The anatomy of the mouth-parts and of the sucking apparatus of some Diptera. Boston : 50 pp., 4 pls.

van Gehuchten, A.
'**90.** Recherches histologiques sur l'appareil digestif de la larve de la *Ptychoptera contaminata*, (part 1). La cellule, vol. 6, p. 185-289, pl. 1-6.

Giles, G. M.
: **02.** A handbook of the gnats or mosquitoes. Edition 2.

de Grandpré, A. D., and de Charmoy, D. d' E.
: **00.** Les moustiques, anatomie et biologie. Contribution à l'étude des Culicidées et principalement des genres Culex et Anopheles de leur rôle dans la propagation de la malaria et de la filariose et des moyens de s' en préserver. Mauritius : 8vo, iv + 59 pp., 5 pls.

Hurst, C. H.
'**90.** The pupal stage of Culex. Studies from biol. lab. Owens college, vol. 2, p. 47-71, pl. 5.
'**96.** The post-embryonic development of a gnat (Culex). Proc. and trans. Liverpool biol. soc., vol. 4, p. 170-191, pl. 5.

Kellogg, V. L.
: **02.** The development and homologues of the mouth parts of insects. Amer. naturalist, vol. 36, p. 683-706, fig. 1-26.

Korscheldt, E., and Heider, K.
'**99.** Textbook of the embryology of invertebrates. English edition ; vol. 3.

Kowalevsky, A.
'**87.** Beiträge zur kenntnis der nachembryonalen entwicklung der Musciden. Zeit. wissensch. zool., vol. 45, p. 542-594, pl. 26-30.

Kräpelin, K.
'**82.** Ueber die mundwerkzeuge der saugenden insecten. Zool. anz., vol. 5, p. 574-576, 3 figs.
'**83.** Zur anatomie und physiologie des rüssels von Musca. Zeit. wissensch. zool., vol. 39, p. 683-719, pl. 40-41.

Lowne, B. T.

'90–'95. The anatomy, physiology, morphology, and development of the blow-fly (*Calliphora erythrocephala*). London: 2 vols., x + 1–350, viii + 351–778 pp., illus.

Macloskie, G.

'88· The poison-apparatus of the mosquito. Amer. naturalist, vol. 22, p. 884–888, 3 figs.

Meinert, F.

'81· Fluernes munddele. Trophi Dipterorum. Copenhagen: 91 pp., 6 pls.

'86· De eucephale myggelarver. Vidensk. selsk. skr., ser. 6, vol. 3, p. 373–493, pl. 1–4.

Miall, L. C.

'93· Dicranota; a carnivorous Tipulid larva. Trans. ent. soc. London, 1893, p. 235–253, pl. 10–13.

Miall, L. C., and Hammond, A. R.

'92· The development of the head of the imago of Chironomus. Trans. Linn. soc. London, ser. 2, zool., vol. 5, p. 265–279, pl. 28–31.

: 00. The structure and life-history of the harlequin fly (Chironomus). Oxford: viii +196 pp., illus.

Miall, L. C., and Shelford, R.

'97· The structure and life-history of *Phalacrocera replicata*. Trans. ent. soc. London, 1897, p. 343–361, pl. 8–11.

Miall, L. C., and Walker, N.

'95· The life-history of *Pericoma canescens* (Psychodidae). Trans. ent. soc. London, 1895, p. 141–153, pl. 3–4.

Nuttall, G. F. H., and Shipley, A. E.

: 01–: 03. Studies in relation to malaria. II. The structure and biology of Anopheles. Journ. of hygiene, vol. 1, p. 46–77, 270–276, 451–484, pl. 1–3, 8–10; vol. 2, p. 58–84; vol. 3, p. 111–215, pl. 6–9.

Pratt, H. S.

'93· Beiträge zur kenntnis der Pupiparen. (Die larve von *Melophagus ovinus*.) Arch. für naturgesch., vol. 59, pt. 1, p. 151–200, pl. 6.

Raschke, E. W.

'87· Die larve von *Culex nemorosus*. Arch. für naturgesch., vol. 53, pt. 1, p. 133–163, pl. 5–6

van Rees, J.

'88· Beiträge zur kenntniss der inneren metamorphose von *Musca vomitoria*. Zool. jahrb., abth. für anat., vol. 3, p. 1–134, pl. 1–2, 10 figs.

Rengel, C.

'96· Ueber die veränderungen des darmepithels bei *Tenebrio molitor* während der metamorphose. Zeit. wissensch. zool., vol. 62, p. 1–60, pl. 1.

Vaney, C.

: 02. Contributions à l'étude des larves et des métamorphoses des Diptères. Ann. de l'univ. Lyon, new ser., fasc. 9, 178 pp., 3 pls.

All figures, unless otherwise specified, are drawn from camera outlines and represent Culex.

*ann*, ring of regenerative cells below esophageal valve.

*ann m*, annular muscle of esophageal valve.

*ant*, antenna.

*ant bd*, antennal bud.

*ant dil*, anterior dorsal dilator muscle of antlia.

*ant hd p*, anterior hard palate.

*ant m*, antennal muscle.

*apo*, apodeme of maxilla (imago).

*apo*, apodeme supporting pharynx wall (larva).

*asc ph*, ascending pharyngeal muscle.

*bc*, buccal cavity.

*b gang*, buccal ganglion.

*blks*, black spot area of larval head.

*bord l*, bordering line of larval head.

*ca*, cardia.

*cae*, caeca of stomach.

*cin*, cingulum muscle.

*co*, colon.

*conv mnd*, converger muscle of mandible.

*cres*, crescentic sclerite of pharynx roof.

*cu*, "cushion" at entrance to pharynx.

*dep li*, depressor of labium.

*dep max*, depressor of the maxilla.

*diag*, diagonal muscle.

*div mnd*, divaricator of the mandible.

*do retr*, double retractor muscle.

*dor dil*, posterior dorsal dilator muscle of antlia.

*dor ph*, dorsal muscles of pharynx.

*el d p*, elevator muscles of dorsal plate.

*el pal m*, elevator muscles of the palate.

*epi*, epipharynx.

*epi m*, epipharyngeal muscles.

*epi tr*, epipharyngeal traverse.

*ext tra*, external trachea.

*fl*, fibers passing from esophagus to shoulder of cardia.

*fl*, flabella.

*fold*, "fold" in front of labial bud.

*f res*, esophageal diverticula.

*gang*, brain.

*h tr*, traverse to hypopharynx and pharynx.

*hyp*, hypopharynx (imago).

*hyp*, hypopharyngeal sclerite (larva).

*hyp m*, hypopharyngeal muscles.

*il*, ileum.

*ilc*, ileo-colon.

*in ant*, inner muscle of antenna.

*int tra*, internal trachea.

*ir fl*, insertion inner retractor flabellae.

*la*, labrum.

*la bd*, labral bud.

*la m*, labral muscle.

*lat dil*, lateral dilator muscles of antlia.

*lat dil oes*, lateral dilator of esophagus.

*lat m*, lateral muscles of pharynx of larva.

*lat ph*, lateral pharyngeal muscles.

*less lat*, lesser lateral muscles.

*li*, labium.

*li n*, labial nerve.

*li tra*, labial trachea.

*m*, circularis muscle of colon (larva).

*mal t*, Malpighian tubules.

*max*, maxilla.
*max-li*, maxillo-labial muscle.
*maxp*, maxillary palpus.
*maxp bd*, bud of maxillary palpus.
*mnd*, mandible.
*mnd m*, mandibular muscle.
*o gang*, optic ganglion.
*oc*, ocellus.
*oes*, esophagus.
*or fl*, insertion outer retractor flabellae.
*ou ant*, outer muscle of antenna.
*pal*, palatum.
*pc*, proboscis canal.
*per*, peritrophic membrane.
*ph*, pharynx.
*prot max*, protractor of the maxilla.
*pump*, antlia.
*r ann*, ring of new cells near anus.
*rec*, rectum.

*retr fl*, retractor muscles of flabellae.
*retr max*, retractor muscle of the maxilla.
*retr ph*, retractors of pharynx.
*s pump*, salivary pump.
*sca*, scythe-shaped thickening on head.
*sd*, salivary duct.
*sg*, salivary glands.
*s-o m*, subocular muscle.
*sph m*, sphincter muscle of antlia.
*st*, stomach or midintestine.
*tent*, tentorium.
*tent m*, tentorial muscle.
*tr*, mental sclerite ("black triangle").
*tr l*, "transverse line".
*ub*, "upper bend" of esophageal valve.
*v*, esophageal valve.
*val m*, valvular muscle.
*vent r ph*, ventral retractors of pharynx.

PLATE 12.

Fig. 1.  Pharynx and salivary pump of female mosquito; seen partly in section and partly in surface view.

Fig. 2.  Semi-diagrammatic section of head of female mosquito, showing pharynx, antlia, and muscles.  The section is supposed to pass near the midline above and anteriorly.  It cuts across the base of the maxillary palp (*mxp*). the "head" of the tentorium of the left side, and then sweeps outward to the origin of the lateral dilator muscles (*lat dil*).  The labium is sectioned along the midline.  Only the proximal ends of the mouthparts are shown.  On the "head" of the tentorium the origin of the inner (*in ant*) and outer muscles of the antenna can be noted.  A portion of the retractor of the maxilla (*retr max*) is removed near the origin of the muscle to display the insertions of the protractor of the maxilla (*prot max*) and the maxillo-labial muscles (*max-li*) on the free end of the maxillary apodeme (*apo*).  The anterior end of the retractor muscles and apodeme are removed to show the insertion of the double retractor muscle (*do retr*) on the articular spur of the maxilla and the attachment of the hypopharyngeal muscle (*hyp m*) to the salivary pump.  The origin end of the hypopharyngeal muscle can be seen on the pharynx wall above the double retractor.  The brain is sectioned, and the buccal ganglion (*b gang*), labial nerve, and one of the rudimentary ocelli (*oc*) are indicated.  The left member of each pair of the elevators of the palate on the roof of the pharynx and of the pair of valvular muscles dorsad to the buccal ganglion at the entrance to the antlia has been cut off, leaving the muscles of the right side in place.  The labral muscles in the clypeus also have been partly removed (*la m*).  Behind the ascending pharyngeal muscles which bind the posterior ends of the pharynx to the vertex of the head (*asc ph*) part of the vestigial anterior dorsal dilator muscle of the antlia is visible. .Below, the lateral pharyngeal muscle can be seen, passing from the outer face of the articular spur of the pharynx to the upper end of the tentorium.  All the mandibular muscle, which runs from the "head" of the tentorium to the mandible, is removed except the insertion end (*mnd m*) and below this can be seen the small maxillary muscle which extends from the wall of the pharynx to the maxilla.  The slender tentorial muscle parallels the nerve cord and runs back into the thorax.  The minute subocular muscle is not figured.

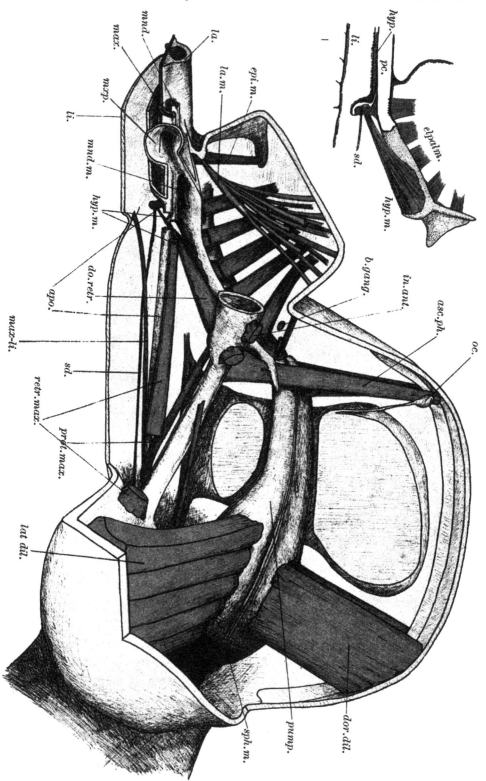

PLATE 13.

Figures 3–12 are sections of the head of a female Culex, each representing an actual section schematized; 3–6 and 9–11 are from one specimen, 7–8 from a second, and 12 is from a third. Figures 13–15 are similar schematized sections of the head of a female Anopheles. The figures are drawn to various scales. The median trachiole of the internal trachea is starred.

Fig. 3. Through bases of mandible and maxilla.

Fig. 4. Insertion of mandibular muscles; second section behind fig. 1.

Fig. 5. Salivary pump and insertion of maxillo-labial muscles; 9th section beyond fig. 1. Sections 10 μ in thickness.

Fig. 6. "Heads" of the tentoria and origins of mandibular muscles; 13th section beyond fig. 1.

Fig. 7. Through posterior spurs of pharynx and the pharyngeo-esophageal valve; specimen B.

Fig. 8. Anterior border of optic ganglion; specimen B.

Fig. 9. Midbrain section of specimen A.

Fig. 10. Origins of retractor maxillae muscles; 39th section beyond fig. 1.

Fig. 11. Through posterior ends of tentoria and insertions of protractors of the maxillae; 41st section beyond fig. 1. Specimen A.

Fig. 12. Region of bulbous portion of antlia; specimen C.

Fig. 13. Section of head of Anopheles; comparable in position to fig. 5.

Fig. 14. Section of head of Anopheles; comparable in position to fig. 11.

Fig. 15. Section of head of Anopheles; comparable in position to fig. 7.

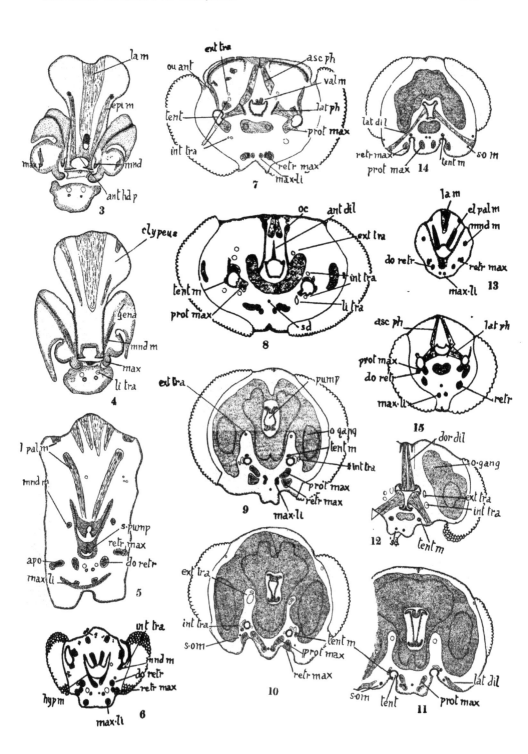

PLATE 14.

Figures 16-19 represent various types of pumping apparatus in the Diptera.

Fig. 16.  Section of head of Anthyomyia ; monantlial type.

Fig. 17.  Section of head of female mosquito ; diantlial and post-neural type.

Fig. 18.  Section of head of Tipulid fly ; diantlial and amphineural type.

Fig. 19.  Head of *Tabanus atratus* ; diantlial and preneural type.   (Free hand.)

Fig. 20.  Alimentary canal of female mosquito with outline of body.  (Free hand.)

Fig. 21.  Longitudinal section through stomach and ileo-colon of female mosquito.

Fig. 22.  Rectum of female mosquito.

Fig. 23.  Salivary gland of a newly emerged imago of *Culex stimulans*.  *ca* = central acinus.

Fig. 24.  Mouthparts of a female pupa in which the hypopharynx (*hyp*) and proboscis canal (*pc*) are forming ; transverse section.

Fig. 25.  Hypopharynx and labium soon after the former is separated from the latter.  Female pupa, transverse section.

Fig. 26.  Alimentary canal of mosquito larva with outline of body.  (Free hand.)  *oes* = esophageal valve.

Fig. 27.  Transverse section of wall of colon of a mosquito larva.

Fig. 28.  Degenerating midintestinal epithelium of a pupa one hour old.  *reg* = regenerative nuclei.

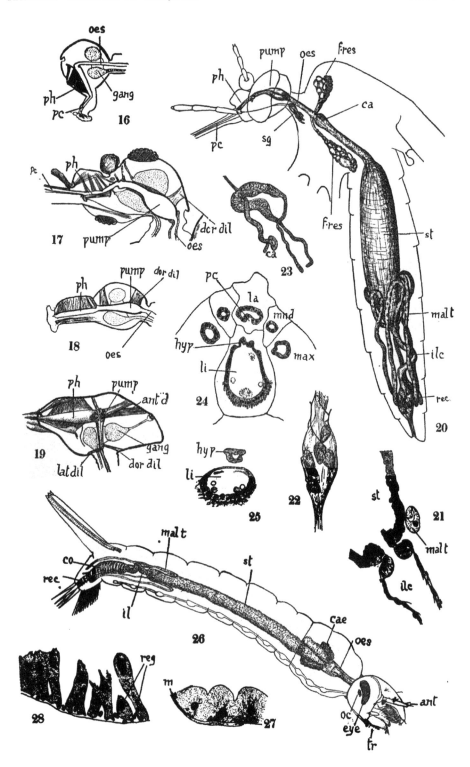

## PLATE 15.

Fig. 29.  Semidiagrammatic view from the ventral aspect of the head of a mosquito larva ; picro-carmine preparation.

Fig. 30.  Similar head from dorsal aspect ; younger specimen.

Fig. 31.  Exuviated head cuticle to show relations of traverses, apodemes of the flabellae, pharynx, epipharynx, etc.

Fig. 32.  Median longitudinal section of head of full grown larva ; compounded several sections.  It is distorted so that the epipharynx lies too far back relatively to the labium and adjacent organs ; (compare with fig. 36) ; labial bud cut in midline between eminences.

Fig. 33.  Dorsal view of pharynx of mosquito larva.

Fig. 34.  Section of lower anterior region of head of mature larva to show relations of labial bud ; bud cut to one side of midline through one of the two eminences.

Fig. 35.  Lateral aspect of larval pharynx, dorsal plate removed.

Fig. 36.  Section of part of head of pupating larva to illustrate fate of ventral fold.  $epi'$ = cuticular sheath of epipharynx.  Compare with figs. 32 and 34.

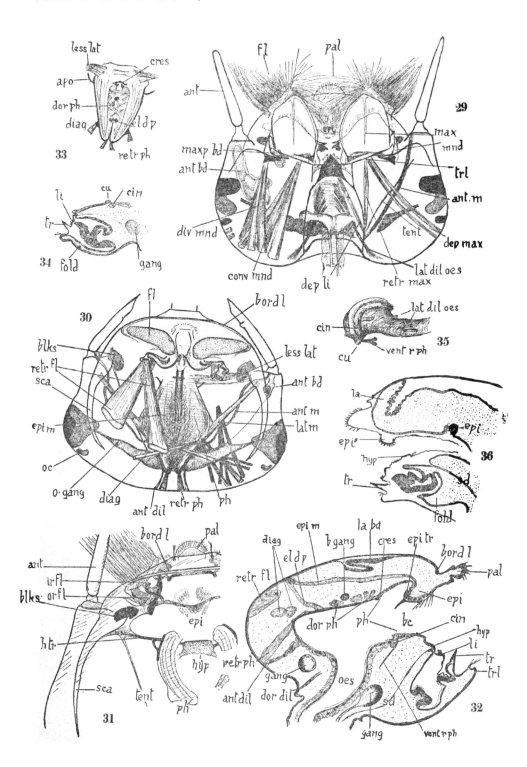

## PLATE 16.

Fig. 37.   Esophageal valve and cardia of a full grown larva, striated border on cardia cells not distinct.

Fig. 38.   Esophageal valve in a pupa four hours old; the left hand wall is ventral.

Fig. 39.   Separation of the degenerated dermis of the epipharyngeal pit (*epi*) from the buccal cavity; longitudinal section of pupa ten hours old. Imaginal muscles (*x*) are beginning to form in the old masses of the retractors of the flabellae and epipharyngeal muscles and the hypopharyngeal muscles (*hyp m*) are forming at the base of the salivary duct.

Fig. 40.   Esophageal valve of a pupa eleven hours old.   The left hand wall is ventral.   *f res* = ventral esophageal diverticulum.

Fig. 41.   Esophageal valve of adult female mosquito; the right hand wall is ventral.

Fig. 42.   Surface of ventral food reservoir or diverticulum of adult female showing the "striae" of the intima, a muscle fiber, and the nuclei of the epithelium.   Bismark brown.

Fig. 43.   Esophageal valve in a pupa ten hours old; the left hand wall is ventral.   *v* = valve, *f res* = ventral diverticulum.

Fig. 44.   Colon and rectum of a larva of the mosquito, longitudinal section.

Fig. 45.   Portion of the wall of the stomach and ileum of a mosquito larva. showing ring of naked epithelium (*x*) at the junction of the two regions where the Malpighian tubules enter.

Fig. 46.   Transverse section of mouthparts of male mosquito.

Fig. 47.   Transverse section of mouthparts of female mosquito; level of tip of palps.   The groove on the dorsal face of the labium in which the other stylets are received, has been obliterated by swelling during dehydration.

Fig. 48.   Ventral aspect of head of Culex pupa at the moment that the larval head cuticle is ruptured.   Dissected out and viewed as an opaque object.   *fold* = fold in front of labial bud (*li*), see pl. 15, figs. 34 and 36.

Fig. 49.   Side view of head of Culex pupa at moment when the respiratory trumpets free themselves by rupture of the thoracic cuticle; dissected out and viewed as an opaque object.

Fig. 50.   Side view of the head of Culex pupa at a time just prior to the complete release of the mouthparts and legs from the larval cuticle.   Their apices were still enclosed in the old sheaths.   Dissected out and viewed as an opaque object.   Drawn on a larger scale than the other figures.

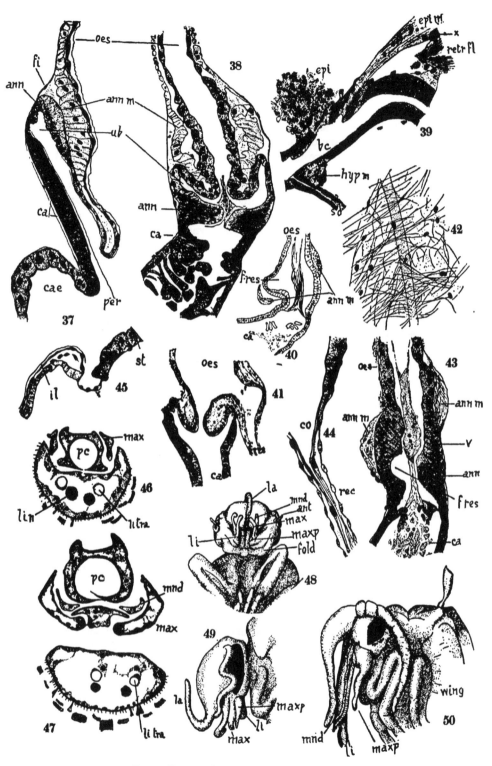

PLATE 17.

Fig. 51. Ileum of pupa twelve hours old, showing the columnar character of the degenerating cells and some reconstructed epithelium (*reg*). *cir m* = circularis muscles.

Fig. 52. Colon and rectum in a pupa five hours old; temporary degeneration of the epithelium of the rectum. Owing to the collapsed walls many cells are cut tangentially and apparently lie in the lumen.

Fig. 53. Same region in a pupa twenty-one hours old, showing the beginnings of a rectal papilla (*p*) and the new-formed ileo-colon (*ilc*). Outside of this lie the remnants of the larval colon (*co*). A few of its muscles (*m*) are still recognizable.

Fig. 54. Diagrammatic longitudinal section of a pupa five hours old.

Fig. 55. Diagrammatic section of a pupa twenty-two hours old. The antlia is not yet formed. *f res* = ventral esophageal pouch.

Fig. 56. Outline of section of head of pupa four hours old. \* = pharyngeo-buccal boundary.

Fig. 57. Outline of section of head of pupa seventeen hours old, epipharynx degenerated, muscles altering. \* = pharyngeo-buccal boundary.

Fig. 58. Outline of section of head of pupa twenty-five hours old, the imaginal muscles and pump perfected. ⟨ = pharyngeo-buccal boundary, *val m = el d p*, *asc ph = retr ph* of figures 56, 57. *epi* = epipharyngeal muscles.

Fig. 59. Invasion of larval colon by cells of the ileum; longitudinal section of pupa seventeen hours old.

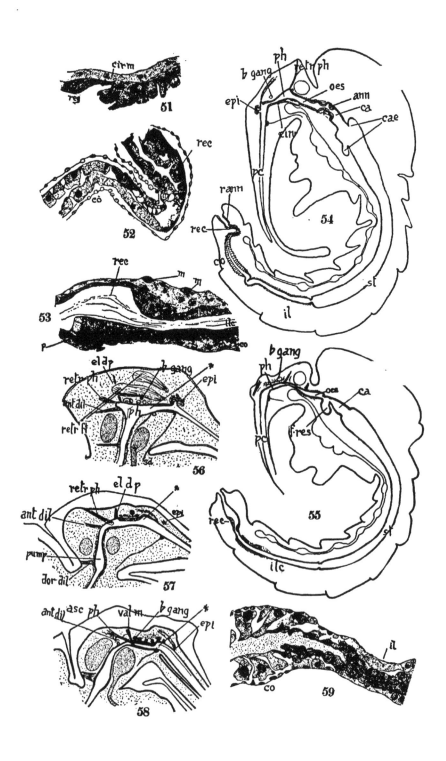

**Proceedings.** 8vo. (For price list of Memoirs, see third page of cover.)

**Vol. 32, No. 4.** Development of Ascus and Spore Formation in Ascomycetes. By J. H. Faull. 37 pp., 5 pls. 50 cts.

No. 3. General anatomy of *Typhlomolge rathbuni*. By E. T. Emerson, 2d. 33 pp., 5 pls. 50 cts.

No. 2. List of shell-bearing Mollusca of Frenchman's Bay, Maine. By D. Blaney. 19 pp., 1 pl. 25 cts.

No. 1. Proceedings of the Annual Meeting, May 4, 1904. 21 pp. 20 cts.

**Vol. 31, No. 10.** The anatomy and development of the terrestrial nemertean (*Geonemertes agricola*) of Bermuda. By W. R. Coe. 40 pp., 3 pls. 50 cts.

No. 9. North American Ustilagineae. By G. P. Clinton. 200 pp. $2.00

No. 8. Pycnogonida collected at Bermuda in the summer of 1903. By L. J. Cole. 14 pp., 8 pls. 25 cts.

No. 7. Trichomes of the root in vascular Cryptogams and Angiosperms. By R. G. Leavitt. 41 pp., 4 pls. 50 cts.

No. 6. Contributions from the Gray herbarium of Harvard university. New series.— No. 27. By B. L. Robinson. 25 pp. 25 cts.

No. 5. Observations on the cytology of *Araiospora pulchra* Thaxter. By C. A. King. 35 pp., 5 pls. 50 cts.

No. 4. The metamorphoses of the hermit crab. By M. T. Thompson. 63 pp., 7 pls. 75 cts.

No. 3. Systematic results of the study of North American land mammals during the years 1901 and 1902. By G. S. Miller, Jr., and J. A. G. Rehn. 85 pp. 50 cts.

No. 2. Proceedings of the Annual Meeting, May 6, 1903. 37 pp. 20 cts.

No. 1. A rare Thalassinid and its larva. By M. T. Thompson. 21 pp., 3 pls. 35 cts.

**Vol. 30, No. 7.** The life history, the normal fission and the reproductive organs of *Planaria maculata*. By W. C. Curtis. 45 pp., 11 plates. 75 cts.

No. 6. Monograph of the Acrasieae. By E. W. Olive. 63 pp., 4 plates. 50 cts.

No. 5. Proceedings of the Annual Meeting, May 7, 1902. 15 pp. 15 cts.

No. 4. Memorial of Professor Alpheus Hyatt. 20 pp. 10 cts.

No. 3. The origin of eskers. By W. O. Crosby. 36 pp., 30 cts.

No. 2. The Medford Dike area. By A. W. G. Wilson. 21 pp., 4 plates 35 cts.

No. 1. Systematic results of the study of North American land mammals to the close of the year 1900. By G. S. Miller, Jr., and J. A. G. Rehn. 352 pp $1.25.

**Vol. 29, No. 18.** The Polychaeta of the Puget Sound Region. By H. P. Johnson. 56 pp., 19 plates. 55 cts.

No. 17. Proceedings of the Annual Meeting, May 1, 1901. 33 pp. 10 cts.

No. 16. Bermudan Echinoderms. A report on observations and collections made in 1899. By H. L. Clark. 7 pp. 10 cts.

No. 15. Echinoderms from Puget Sound : Observations made on the Echinoderms collected by the parties from Columbia University, in Puget Sound in 1896 and 1897. By H. L. Clark. 15 pp., 4 plates. 30 cts.

No. 14. Glacial erosion in France, Switzerland and Norway. By William Morris Davis. 50 pp., 8 plates. 50 cts.

No. 13. The embryonic history of imaginal discs in Melophagus ovinus L., together with an account of the earlier stages in the development of the insect. By H. S. Pratt. 32 pp., 7 plates. 75 cts.

No. 12. Proceedings of the Annual Meeting, May 2, 1900. 18 pp. 10 cts.

No. 11. A revision of the systematic names employed by writers on the morphology of the Acmaeidae. By M. A. Willcox. 6 pp. 10 cts.

No. 10. On a hitherto unrecognized form of blood circulation without capillaries in the organs of vertebrata. By Charles Sedgwick Minot. 31 pp. 35 cts.

No. 9. The occurrence of fossils in the Roxbury conglomerate. By Henry T. Burr and Robert E. Burke. 6 pp., 1 plate. 20 cts.

No. 8. The blood vessels of the heart in Carcharias, Raja, and Amia. By G. H. Parker and F. K. Davis. 16 pp., 3 plates. 25 cts.

No. 7. List of marine mollusca of Coldspring Harbor, Long Island, with descriptions of one new genus and two new species of Nudibranchs. By Francis Noyes Balch. 30 pp., 1 plate. 35 cts.

No. 6. The development of Penilia schmackeri Richard. By Mervin T. Sudler. 23 pp., 3 plates. 30 cts.

CPSIA information can be obtained
at www.ICGtesting.com
Printed in the USA
BVHW090439201118
533516BV00014B/807/P